PRACTICAL ASPECTS
OF
ION TRAP MASS
SPECTROMETRY

Volume II

*Ion Trap
Instrumentation*

MODERN MASS SPECTROMETRY

A Series of Monographs on
Mass Spectrometry and Its Applications

EDITOR-IN-CHIEF
Thomas Cairns, Ph.D., D.Sc.

Forensic Applications of Mass Spectrometry
Edited by Jehuda Yinon

Practical Aspects of Ion Trap Mass Spectrometry, Volume I
Fundamentals of Ion Trap Mass Spectrometry
Edited by Raymond E. March and John F. J. Todd

Practical Aspects of Ion Trap Mass Spectrometry, Volume II
Ion Trap Instrumentation
Edited by Raymond E. March and John F. J. Todd

Practical Aspects of Ion Trap Mass Spectrometry, Volume III
Chemical, Environmental, and Biomedical Applications
Edited by Raymond E. March and John F. J. Todd

PRACTICAL ASPECTS OF ION TRAP MASS SPECTROMETRY

Volume II

Ion Trap Instrumentation

Edited by

Raymond E. March
John F. J. Todd

CRC Press

Boca Raton   New York   London   Tokyo

Library of Congress Cataloging-in-Publication Data

Practical aspects of ion trap mass spectrometry / edited by Raymond E. March, John F. J. Todd

    p. cm. — (Modern mass spectrometry)
    Includes bibliographical references and index.
    ISBN 0-8493-4452-2 (vol. 1)
    ISBN 0-8493-8253-X (vol. 2)
    ISBN 0-8493-8251-3 (vol. 3)
    1. Mass spectrometry.   I. March, Raymond E.   II. Todd, John F. J.   III. Series
QD96.M3P715   1995
539.7′028′7—dc20
                                         95-14146
                                           CIP

DEDICATION

To ion trappers, young and old, everywhere.

FOREWORD

Publication of the volumes in the new series entitled *Modern Mass Spectrometry* represents a milestone for scientists intimately involved with the practice of mass spectrometry in all its various forms. Forthcoming monographs in this series will focus on various selected topics within the rapidly expanding realm of mass spectrometry. Individual volumes will provide in-depth reports on mainstream developments where there is an urgent need for a specific mass spectrometry treatise in an active and popular area.

While mass spectrometry as a field is quite well served by several publications and a number of societies, the application of mass spectrometric techniques across the basic scientific disciplines has not yet been recognized by existing journals. The present distribution of research and application papers in the scientific literature is widespread. There is a multi-disciplinary audience requiring access to concise reports illustrating the latest successful approaches to difficult analytical problems.

The distinguished members of the Editorial Advisory Board all agreed that a platform exists for a premier book series with high standards to cover comprehensively general aspects of developing mass spectrometry. Contributing authors to the series will provide concise reports together with a bibliography of publications of importance, selecting worthy examples for inclusion. Due to the rapid and extensive growth of the literature in mass spectrometry, there is a need for such reports by authorities who critique the entire subject area. There is an increasing urgency to provide readers with timely, informative, and cogent reviews stripped of outdated material.

I believe that the decision to publish the series *Modern Mass Spectrometry* reflects the realization that increasing numbers of mass spectrometrists are applying nascent state of the art approaches to some fascinating problems. Our challenge is to develop the best forum in which to present these emerging issues to encourage and stimulate other scientists to comprehend and adopt similar strategies for other projects.

Lofty ideals aside, our immediate goal is to build a reputable series with scientific authority and credibility. To this end we have assembled an excellent Editorial Advisory Board that mirrors the prerequisite multi-disciplinary exposure required for success.

I am confident that the mass spectrometry community will be pleased with this "new publication" approach and will welcome the opportunities it presents to foster the development of interactions between the various scientific disciplines. No doubt there is a long and difficult road ahead of us to ensure that the series grows into a position of leadership, but I am convinced that the hard work of our outstanding Editorial Advisory Board, and the enthusiasm of our authors and readers, will achieve the degree of success for which we all seek.

Thomas Cairns
Editor-in-Chief

PREFACE

This monograph is Volume II in the series *Modern Mass Spectrometry*, published by CRC Press; it is part of a mini-series of three volumes on *Quadrupole Ion Trap Mass Spectrometry*. Volume II, which is entitled "Instrumentation", is a companion to Volume I, entitled "Fundamentals".

The Ion trapping field continues to be a very active one and, as we acknowledged in the Preface to Volume I, we are extremely grateful to many of the principal players in the field who, in contributing to this two-volume set, have enabled us to create a concise and coherent account of the state of ion trap mass spectrometry. While it was a great privilege for us to be invited, in 1991, to undertake the preparation of an initial monograph on the quadrupole ion trap for CRC Press, it is to the enormous credit of our contributors that, in responding so positively to our invitations to participate in this endeavor, we shall be able to prepare three volumes in all. The "Topsy-like" growth of this mini-series on the quadrupole ion trap has been described in the Preface to Volume I. Following publication of this monograph, a third volume on the ion trap, entitled "Chemical, Biomedical, and Environmental Applications" is to appear.

While the history of the quadrupole ion trap has been presented in tabular form as "The Ages of the Ion Trap", in Chapter 2 of Volume I, it would be remiss of us were we not to comment on two landmarks in this history. First, the invention of the ion trap by Wolfgang Paul and Hans Steinwedel, and which was recognized by the award of the 1989 Nobel Prize in Physics, in part, to Wolfgang Paul and Hans Dehmelt. Second, the discovery announced in 1983 of the mass-selective instability scan by George C. Stafford, Jr. On these two landmarks rests the entire field of ion trap mass spectrometry. The award of the Nobel Prize, which recognized the discovery and utilization of the ion trap, has served to focus attention more sharply on the ion trapping field, and has been responsible, though indirectly, for the commissioning of this mini-series. However, George Stafford's discovery made possible the commercial development of a powerful and versatile ion trap mass spectrometer and, in so doing, made the several variations of this instrument readily available to scientists in a wide variety of fields.

Volume I was presented in two parts, each composed of four chapters. In Part 1, entitled Fundamentals, accounts of the development and theory of the quadrupole ion trap and of its utilization as an ion storage device were followed by a description of nonlinear ion traps and an account of the commercialization of the quadrupole ion trap. Part 2 entitled Ion Activation and Ion/Molecule Reactions, focused on the environment within the ion trap; that is, the movement of ions within the trap, the modifications of ion motion by repeated collisions with helium buffer gas atoms, ion activation, and on those collisions of ions with molecules that lead to chemical change.

The principal objective of this monograph is to convey to the reader an appreciation of the ion trap as an instrument whose versatility is incompletely known, yet it can be used in many avenues of research and in many applications often in tandem with other instruments or components, such as external ion sources and lasers. This volume is composed of eight chapters and is presented in five parts, namely Enhancement of Ion Trap Performance, Ion Trap Confinement of Externally Generated Ions, Ion Structure Differentiation in an Ion Trap, Lasers and the Ion Trap, and Ion Traps in the Study of Physics.

PART 1: ENHANCEMENT OF THE ION TRAP PERFORMANCE.

Chapter 1 is devoted to a discussion of two related areas which have been opened up recently: those of High Resolution Mass Spectrometry and Mass Measurement Accuracy. The authors describe the results of recent researches in these areas which have enormous implications for future application of the ion trap as a mass spectrometer of high resolution and accurate mass assignment.

PART 2: ION TRAP CONFINEMENT OF EXTERNALLY GENERATED IONS.

The confinement of the ion trap of ions generated externally clearly makes possible utilization of the ion trap with a variety of external ion sources. Chapters 2 and 3 are complementary in that Chapter 2 deals with the confinement of ions with high initial kinetic energy, such as are formed in a sector mass spectrometer; and Chapter 3 discusses the confinement of ions with relatively low kinetic energy, that is, ions emerging from an electrospray source. Both achievements are quite remarkable. These chapters show that trajectories of high kinetic energy ions can be controlled so that such ions are confined, while Chapter 3 shows that multiply protonated species also may be confined and, following isolation, individual charge states may be studied.

PART 3: ION STRUCTURE DIFFERENTIATION IN AN ION TRAP.
Chapter 4 deals with an entirely new aspect of ion trap behavior in which the dipole moment of a gaseous ion appears to interact subtlely with the trapping field. As a result of this interaction, the displacement of the experimental value from the anticipated value of the mass/charge ratio, as determined by ion ejection, varies according to the polarizability of the parent neutral.

PART 4: LASERS AND THE ION TRAP.
Chapters 5 and 6 are devoted to accounts of the application of lasers to the study of trapped ions. Chapter 5 concentrates upon the laser photodissociation of gaseous ions confined within the ion trap. Both single photon and multiphoton modes are discussed. Complementarily, Chapter 6 is concerned with laser desorption of gaseous ions from solid, that is, involatile, samples, and with matrix-assisted laser desorption ionization. This chapter is presented in two parts, the first is concerned with laser desorption, and the second extends this discussion to matrix-assisted laser desorption ionization using tandem mass spectrometry.

PART 5: ION TRAPS IN THE STUDY OF PHYSICS.
Chapters 7 and 8 are devoted to Physics applications of the ion trap. Chapter 7 concentrates almost entirely on the Penning trap, the practical realization of the Penning trap, ion cooling within the Penning trap, and the detection of resonances. The Penning trap theme continues into Chapter 8. This chapter treats the Paul trap as a collection device for the accumulation of ions and compaction of ion clouds preparatory to transfering the ion cloud to a Penning trap for high resolution spectroscopy. Ion clouds are treated as collections of particles which can be described in terms of action diagrams.

It is noteworthy that nine of the authors contributing to this two-volume set are, or were until recently, graduate students in the contributing laboratories. It is our belief that this figure supports our contention that research in the ion trapping field is attractive and challenging for graduate students which, we hope, augurs well for the future. With the advent of desk-top computer-controlled high performance instruments, such as the commercial version of the ion trap, a much greater fraction of graduate student time can be spent interpreting results and designing new experiments rather than constructing or repairing laboratory-made apparatuses.

We believe that the breadth of the contributions in this monograph illustrates clearly both the wide variety of research that has been recently undertaken and the enormous possibilities for future ion trap research. In support of this hypothesis, we intend to follow Volume II with a further monograph dealing with applications in the chemical, biomedical and environmental fields.

Finally we wish to thank the many people who have assisted us in one way or another with the many tasks that make up completion of a manuscript. First of all, to our contributors without whom this monograph would not have appeared, we give thanks for their individual inspirations. We thank them for the fruits of their labors, and for their patient toleration. We thank the staff at CRC Press for their ready cooperation and encouragement, and the Series Editors, Dr. Thomas Cairns and Dr. M. Allen Northrup for their guidance and assistance. In the Trent laboratory, REM thanks Jenny Bazdikian, Oscar Vega, Jeff Plomley, Peter Popp, and Dr. Richard Hughes for their enormous contribution in proof-reading and checking manuscripts. We also thank Bonnie Mackinnon for her ever-ready assistance with a myriad of word-processing minutiae.

Raymond E. March
John F. J. Todd

THE EDITORS

Raymond E. March, Ph.D., is presently Professor of Chemistry at Trent University in Peterborough, Ontario, Canada. He obtained his B.Sc. degree from the University of Leeds in 1957; his Ph.D. degree, which he received in 1961 from the University of Toronto, was supervised by Professor John C. Polanyi.

Dr. March has conducted independent research for over 28 years and has directed research in gas phase kinetics, optical spectroscopy, gaseous ion kinetics, analytical chemistry, and mass spectrometry.

He has published and/or coauthored over 130 scientific papers in the above areas of research with emphasis on mass spectrometry, both with sector instruments and quadrupole ion traps. Dr. March is a coauthor with Dr. Richard J. Hughes of *Quadrupole Storage Mass Spectrometry*, published in 1989.

Professor March is actively engaged in the supervision of graduate student research and is an Adjunct Professor of Chemistry at Queen's and York Universities in Ontario; he is the Associate Director of the Trent/Queen's Cooperative Graduate Program. Professor March is a Fellow of the Chemical Institute of Canada and a member of the American, British, and Canadian Societies for Mass Spectrometry. Research in Dr. March's laboratory is supported by the Natural Sciences and Engineering Research Council of Canada, the Ontario Ministry of the Environment, and Varian Associates. Dr. March has enjoyed long-term collaborations with the co-editor, John Todd, and with colleagues at the University of Provence and Pierre and Marie Curie University in France, and with colleagues in Italy.

John F. J. Todd, Ph.D., is currently Professor of Mass Spectroscopy and was, until recently, Director of the Chemical Laboratory at the University of Kent, Canterbury, England. He obtained his B.Sc. degree in 1959 from the University of Leeds, from whence he also gained his Ph.D. degree and was awarded the J. B. Cohen Prize in 1963, working in the radiation chemistry group led by Professor F. S. (now Lord) Dainton, FRS. He was a postdoctoral research fellow in the laboratory of the late Professor Richard Wolfgang at Yale University, Connecticut, from 1963 through

1965 and was one of the first appointees to the academic staff of the, then new, University of Kent in 1965.

Since arriving in Canterbury Professor Todd's research interests have encompassed mass spectral fragmentation studies, gas discharge chemistry, ion mobility spectroscopy, analytical chemistry, and ion trap mass spectrometry. His work on ion traps commenced in 1968, and he first coined the name QUISTOR for quadrupole ion store, to describe a trap coupled to a quadrupole mass filter for external mass analysis. Acting as a consultant to Finnegan MAT, he was a member of the original team that developed the ion trap commercially. Research in Professor Todd's laboratory has been supported by the Science and Engineering Research Council, the Engineering and Physical Sciences Research Council, the Defense Research Agency, the Chemical and Biological Defence Establishment, and Finnegan MAT, Ltd.

Dr. Todd has published and/or co-authored over 100 scientific papers, concentrating on various aspects of mass spectrometry. With Dr. Dennis Price, he co-edited several volumes of *Dynamic Mass Spectrometry,;* he edited *Advances in Mass Spectrometry 1985;* and he is currently editor of the *International Journal of Mass Spectrometry* and *Ion Processes.*

Professor Todd is actively engaged in the supervision of full-time and industrial-based part-time graduate students, all working in the field of mass spectrometry. He is a Chartered Chemist and a Chartered Engineer and is currently an elected member of the Council of the Royal Society of Chemistry. He recently completed a four-year term as Treasurer of the British Mass Spectrometry Society. Outside the immediate confines of academic work, he was for ten years Master of Rutherford College at the University of Kent and also for four years was appointed as the Chairman of the Canterbury and Thanet Health Authority. He is presently a Governor of the Clergy Orphan Corporation and of Canterbury Christ Church College. He has enjoyed long-term collaborations with co-editor Professor Raymond March, with colleagues at Finnigan MAT in the United Kingdom and the United States and with groups in Nice (France) and Padova and Torino (Italy).

CONTRIBUTORS

Olga Bortolini
Servizio di Spettrometria di Massa del
CNR
Padova, Italy

Jennifer Brodbelt
Department of Chemistry
University of Texas
Austin, Texas

R. Graham Cooks
Henry Bohn Hass Distinguished Professor
Chemistry Department
BRWN Laboratory
Purdue University
West Lafayette, Indiana

Kathleen A. Cox
The Center for Mass Spectrometry
Department of Chemistry
UMIST
Manchester, United Kingdom

Gary L. Glish
Department of Chemistry
University of North Carolina
Chapel Hill, North Carolina

Peter Kofel
Institut für Organische Chemie
Universität Bern
Bern, Switzerland

M. David Lunney
Foster Radiation Laboratory
McGill University
Montreal, Quebec
Canada

Raymond E. March
Department of Chemistry
Trent University
Peterborough, Ontario
Canada

Scott A. McLuckey
Chemical and Analytical Sciences Division
Oak Ridge National Laboratory
Oak Ridge, Tennessee

Robert B. Moore
Foster Radiation Laboratory
McGill University
Montreal, Quebec
Canada

Jae C. Schwartz
Finnigan MAT
San Jose, California

James L. Stephenson
Department of Chemistry
University of Florida
Gainesville, Florida

John F. J. Todd
Chemical Laboratory
University of Kent
Canterbury, Kent
England

Pietro Traldi
Consiglio Nazionale Ricerche
Area di Ricerca Di Padova
Servizio di Spettrometria di Massa
Padova, Italy

Gary J. Van Berkel
Chemical and Analytical Sciences Division
Oak Ridge National Laboratory
Oak Ridge, Tennessee

Fernande Vedel
Physique des Interactions
Ioniques et Moléculaires
Université de Provence
Marseille, France

Gunther Werth
Institut fur Physik
Universitat-Mainz
Mainz, Germany

Jon D. Williams
Kraft General Foods, Inc.
Glenview, Illinois

Richard A. Yost
Department of Chemistry
University of Florida
Gainesville, Florida

TABLE OF CONTENTS

Part 1. Enhancement of Ion Trap Performance

Chapter 1
High Mass, High Resolution Mass Spectrometry 3
Jon D. Williams, Kathleen A. Cox, Jae C. Schwartz, and R. Graham Cooks

Part 2. Ion Trap Confinement of Externally-Generated Ions

Chapter 2
Injection of Mass-Selected Ions into the Radiofrequency Ion Trap 51
Peter Kofel

Chapter 3
Electrospray and the Quadrupole Ion Trap 89
Scott A. McLuckey, Gary J. Van Berkel, Gary L. Glish, and Jae C. Schwartz

Part 3. Ion Structure Differentiation in an Ion Trap

Chapter 4
Evaluation of the Polarizability of Gaseous Ions 145
Olga Bortolini and Pietro Traldi

Part 4. Lasers and the Ion Trap

Chapter 5
Photodissociation in the Ion Trap.. 163
James L. Stephenson, Jr. and Richard A. Yost

Chapter 6
Laser Desorption in a Quadrupole Ion Trap 205
Jennifer S. Brodbelt, Rafael R. Vargas, and Richard A. Yost

Part 5. Ion Traps in the Study of Physics

Chapter 7
High Precision Mass Spectrometry in the Penning Trap 237
Fernande Vedel and Gunther Werth

Chapter 8
The Paul Trap as a Collection Device 263
Robert B. Moore and M. David Lunney

Author Index 303
Chemical Index..305
Subject Index...309

PART 1

Enhancement of
Ion Trap Performance

Chapter 1

HIGH MASS, HIGH
RESOLUTION MASS
SPECTROMETRY

Jon D. Williams, Kathleen A. Cox, Jae C. Schwartz, and R. Graham Cooks

CONTENTS

I. Introduction 4

II. Experimental 6

III. Extension of the Mass/Charge Range 8
   A. Introduction 8
   B. Reduced Size Electrodes 11
   C. Reduction of the RF Drive Frequency 12
   D. Resonance Ejection 14
   E. Future 21

IV. High Resolution 21
   A. Introduction 21
   B. Offset-DAC 21
   C. Mass Spectra 24
      1. Applications 24
      2. Simulations 27
   D. Disadvantages 28

0-8493-4452-2/95/$0.00+$.50
© 1995 by CRC Press, Inc.

V.   High Resolution in Tandem (MS/MS) Experiments 31
      A.   Isolation 31
      B.   Applications 33
      C.   Future 35

VI.   Mass Measurement Accuracy 35
      A.   Introduction 35
      B.   Internal Calibration of Low Resolution Mass Spectra . . 37
      C.   Calibration of Higher Resolution Mass Spectra 38
      D.   Calibration of Higher Resolution Tandem
            Mass Spectra 40
      E.   Future 43

VII. Conclusion 44

Acknowledgments 45

References 46

I. INTRODUCTION

The ability to analyze and detect high mass compounds with high mass resolution and high mass measurement accuracy is becoming increasingly important. With a recent sequence of developments in ionization techniques progressing from plasma desorption[1] through secondary-ion mass spectrometry,[2] fast-atom bombardment,[3] matrix-assisted laser desorption,[4] and electrospray,[5] mass spectrometry has gained prominence as a technique for obtaining information on biomolecules with molecular weights > 1000 Da. The development of high performance capabilities and their utilization for peptide and protein analysis have been well documented[6] for sector, quadrupole, time-of-flight (TOF), hybrid and Fourier transform-ion cyclotron resonance (FT-ICR) mass spectrometers. This chapter describes the development and the utilization of such capabilities for the quadrupole ion trap mass spectrometer (ITMS).

Typically, sector-field mass analyzers are used for high mass as well as high resolution measurements, with detection of ions with ratios in excess of 10^5 Da per charge, resolutions well in excess of 100,000 (full width at half-maximum (FWHM) definition), and mass measurement accuracies better than 1 ppm being achieved.[7,8] Time-of-flight mass analyzers which separate ions of equal charge and energy on the basis of velocity, are capable of m/z detection up to the range 10^5 to 10^6 Da per charge. Mass resolutions of < 4000 (FWHM) for biomolecules are usual, although special systems have resulted in resolutions up to 28,000.[9] Mass measurement accuracies for TOF experiments, however, are typically not much better than 0.1%. Fourier-transform ion cyclotron resonance mass spectrometry distinguishes ions on the basis of their characteristic cyclotron frequencies. Mass/charge ratios of 3×10^5 have been recorded.[10] Resolution varies directly with field strength and extremely high resolutions can be achieved at low mass; a resolving power exceeding 10^6 for multiply charged ions at 8.6 kDa has been observed.[11]

Originally, the quadrupole ion trap was introduced commercially as a simple gas chromatography (GC) detector.[12,13] It was not intended to be a high performance mass spectrometer. However, its compact size and high sensitivity led us to embark upon a research program to transform the ion trap into a high performance mass spectrometer, capable of analyzing both low and high mass ions, and of achieving high mass resolution and mass accuracy. Capabilities to perform tandem mass spectrometry on low mass species were established[14] through radiofrequency/direct current (RF/DC) isolation of precursor ions and resonant excitation with a dipole field of appropriate frequency, and the ability to inject ions from an external ion source was achieved.[15] These experiments paved the way for the use of thermospray,[16,17] electrospray,[18–20] ionspray,[21] secondary ion mass spectrometry,[22] and laser desorption[15,23–25] to analyze thermally labile compounds using the ion trap.

Techniques to extend the m/z range of the trap were the next improvements to be developed.[22,26,27] Three methods for mass range extension were explored: reduction of the dimensions of the trap,[26,27] reduction of the main RF drive frequency,[27] and resonant ion ejection using an alternating current (AC) signal of appropriate frequency, applied across the end-cap electrodes during the mass-analysis scan.[22,27] Using resonant ejection with a modest reduction of the drive frequency, the m/z range was extended up to 72,000 Da per charge.[27] The mass resolution was observed to increase by slowing the rate of mass-selective instability when resonant ejection was performed.[27] High mass resolution was achieved by using very slow scans.[28–32] A resolution in excess of 10^6 has been demonstrated on a CsI cluster ion[31] using very slow scans, and resolutions in excess of 10^7 for m/z 614 from the calibration compound perfluorotributylamine (FC-43) using even slower scan rates have been reported.[33] However, low mass measurement accuracy limits the value of this approach. High resolution capabilities have been successfully utilized on the quadrupole ITMS in mass spectral studies of molecular ions as well as in tandem mass spectrometry (MS^n) experiments. These experiments have included isobaric separation of precursor and fragment ions for structural elucidation as well as determination of charge states of ions.[28,30–32] High resolution MS/MS experiments have been performed also on singly[32] and multiply charged[28,30] peptide ions in order to elucidate fragmentation mechanisms.

In order to utilize the mass range extension and high mass resolution capabilities to the fullest extent, a calibration method has been developed so that accurate mass assignments are possible.[34] The use of resonance ejection for mass range extension results in slight shifts (which originate from a variety of sources) from the predicted mass assignments. Calibration can be performed externally, injecting salt cluster ions under the same conditions as the analyte ions and constructing a calibration curve of the observed mass shifts vs. the theoretical masses, and this method gives mass accuracies of approximately 0.1 percent.[27] To improve mass measurement further, a split probe tip has been designed to allow calibrant and analyte ions to be physically separated on the probe to prevent quenching of analyte signal. Co-injection of analyte and calibrant into the trap gives the advantages of an internal standard for mass calibration of the instrument.[34] The co-injection method becomes extremely important in higher resolution experiments where the behavior of ions in the trap (and thus their experimentally observed masses) are highly dependent on the precise experimental conditions including helium pressure and ion density. For higher resolution MS and MS/MS experiments, a peak-matching routine has been developed to allow even greater mass measurement accuracies.[34] Mass measurements within ±0.05 Da can be made throughout the entire calibrated m/z range (up to 1500 Da per charge) when internal calibrants and peak matching are used.[35]

This chapter examines the techniques used to achieve high mass detection, high mass resolution, and high mass measurement accuracy. Comparisons are made between each of the methods used to extend the m/z range of the ITMS. The theory and operation of the offset DAC, which is used to slow the mass analysis scan in order to improve mass resolution, is described in detail, and examples of high resolution MS and MS/MS spectra are shown. Finally, the problems associated with achieving high mass measurement accuracy for both low and high resolution experiments are discussed, together with the methods being used to minimize their effects.

II. EXPERIMENTAL

Quadrupole traps fitted with external ion sources (Cs⁺ SIMS,[22] electrospray,[28] or laser desorption[23]) have been described previously. Instruments so equipped have the capabilities to detect ions with high m/z ratios and to perform high resolution MS and MS/MS experiments. Other typical modifications made to these instruments include:

1. An off-axis dynode detection system
2. An amplifier installed to increase the maximum voltage output of the supplementary AC signal applied to the end-caps from 3 to 24 V_{0-p}

3. A circuit to attenuate the output of the 12-bit accuracy mass control DAC and to sum the resulting voltage with the output from either the 16-bit DAC that controls the DC voltage applied to the ring electrode or an external analog voltage source (for mass measurement accuracy experiments)

4. Removal of the output from the DC selective storage box from the ring electrode ($a_z = 0$ for all experiments)

The RF drive frequency for the ion trap at Purdue University was 1.1 MHz, unless otherwise noted, while the ion trap at Finnigan MAT operated at a drive frequency of 0.8803 MHz.[28] Both instruments used the standard ring and end-cap electrodes which form the stretched geometry trap provided with the commercial ITMS except for the experiments to evaluate the performance of reduced size traps.[26,27] The reduced size electrodes, which were designed to the theoretical quadrupole geometry, were machined to tolerances of 0.0001 in. to calculated surface profiles using numerically controlled machine tools at Los Alamos National Laboratory.[26,27]

Conditions for Cs^+ SIMS experiments on peptides are listed in Table 1. The conditions for electrospray experiments are found elsewhere.[28,30] In the case of CsI or RbI samples, 1 μl of an aqueous sample was applied on the probe tip and was dried *in vacuo*. In peptide analyses, 1 μl of a glycerol:thioglycerol mixture was applied to the probe tip along with 1 μl of peptide dissolved in a 1:1 $CH_3OH:H_2O$ solution. Typically, low picomole to femtomole amounts of peptide were examined.

A timing diagram used to obtain mass spectra of high mass species by resonant ejection is shown in Figure 1. Note the similarity of this scan function to those used to obtain mass spectra of low mass ions. When Cs^+ SIMS is used, the potential of the Cs^+ extraction lens is set to allow primary ions to reach the probe tip during the first stage of the timing sequence. Sample ions are desorbed from the probe tip and are injected and subsequently trapped. The amplitude of the RF applied to the ring electrode upon injection is optimized for each experiment. In the final portion of the scan, the RF amplitude is ramped to mass analyze the injected ions using the mass-selective instability scan.[13] As shown in Figure 1, a supplementary AC voltage is applied to the end-caps during the mass scan to cause ions to be ejected resonantly from the trap. In addition to use of the supplementary AC voltage to achieve mass range extension, reduction of the dimensions of the trap and reduction of the RF drive frequency also were examined. For higher resolution MS experiments, the AC modulation voltage was applied to the end-caps, and the rate of the mass-selective instability scan was slowed by using the offset DAC/attenuator circuit.[27] Tandem mass spectrometry experiments required additional stages to isolate precursor ions using reverse-then-forward scans[32] and to fragment these ions by resonant excitation before the mass scan

TABLE 1

Typical Ion Trap Operating Parameters for the Analysis of Peptides

Cesium Desorption Source

Source voltage	7 kV (float)
Primary ion current:	2.25 mA
Gate on:	–375 V (relative to Cs⁺ high voltage)
Gate off:	+75 V (relative to Cs⁺ high voltage)
Grid:	–600 V (relative to Cs⁺ high voltage)

Injection and Ionization System

Ion volume:	4–15 V
Extractor:	10–20 V
Ionization time:	20 ms
Sample size:	10 pmol
Matrix:	Glycerol:thioglycerol (1:1)
q_z:	0.06

MS Parameters

Ejection voltage	7.5 V_{0-p}
Mass range extension:	2–10
RF frequency:	1.1 MHz
r_0:	1.0 cm

MS/MS Parameters

Helium pressure:	0.5–1 mtorr
Activation time:	30 ms
Activation frequency:	>78 kHz (q_z = 0.2)
Activation voltage:	1–2 V_{0-p}

(Adapted from Cox, K.A. et al., *Biological Mass Spectrometry*, vol. 21. John Wiley and Sons, 1992, 226.)

was performed. The convention adopted in presenting high resolution results is to use the mass of the known $^{12}C^1H^{14}N^{16}O$ isotopic species and to round down in order to count the spectrum in integral steps.

III. EXTENSION OF THE MASS/CHARGE RANGE[22,26,27]

A. Introduction

To understand how high mass ions can be mass analyzed, it is most instructive to refer to the stability diagram (Figure 2). The most important characteristic of the trajectory of an ion is whether the trajectory is stable or unstable within the confines of the ion trap; when the q_z, a_z Mathieu parameters fall within the stable region of the stability diagram (Figure 2), the trajectory can be stable, subject to some constraints on the

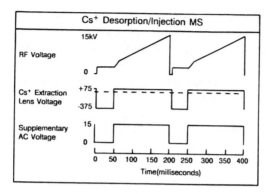

FIGURE 1

Ion trap scan program used to record mass spectra generated by Cs' SIMS. (Adapted from Kaiser, R.E. et al., *Int. J. Mass Spectrom. Ion Proc.*, 106, 79, 1991.)

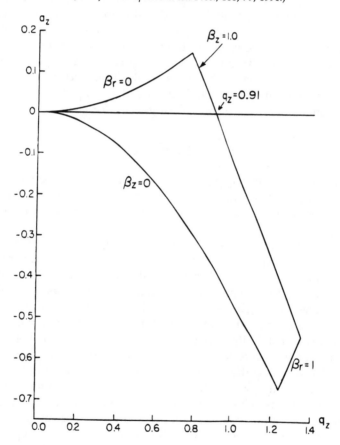

FIGURE 2

Stability diagram denoting β_r and β_z instability boundaries.

initial velocity and position. Outside this region, the trajectory is unstable regardless of the initial velocity and position. An ion must have a stable trajectory in both the r- (radial) and z- (axial) directions to remain in the trap.

The parameterized coordinates of the stability diagram, as expressed by the general representation of the Mathieu equation coefficients[36] are

$$q_z = \frac{8eV}{m(r_0^2 + 2z_0^2)\Omega_0^2} \tag{1}$$

and

$$a_z = \frac{-16eU}{m(r_0^2 + 2z_0^2)\Omega_0^2} \tag{2}$$

where V is the peak RF voltage (ranging from 0 to 7500 V_{0-p} in the Finnigan ion trap), U is the DC potential applied to the ring electrode (often zero), r_0 is the inscribed radius of the ring electrode (1 cm), z_0 is the actual measured distance from the center of the trapping cavity to the apex of the end-cap electrode (0.783 cm), Ω_0 is the angular drive frequency ($\Omega_0/2\pi$ measures 1.1 MHz), m is the mass of the ion, and e is its charge. Rearrangement of the terms that define the Mathieu parameter q_z creates the following expression for mass/charge m/e (or m/z):

$$\frac{m}{z} = \frac{8V}{(r_0^2 + 2z_0^2)\,\Omega_0^2 q_z} \tag{3}$$

When mass analysis is performed using the mass-selective instability scan, $q_z(q_{eject}) = 0.908$ and $a_z = 0$, the commercial Finnigan MAT ion trap can mass analyze ions with m/z ratios up to approximately 650 Da per charge.

From inspection of this equation, it is possible to increase the m/z range of the ion trap in four ways: (1) by increasing the maximum RF voltage applied to the ring electrode, a step which has not been taken; (2) by decreasing the dimensions of the trap; (3) by decreasing the RF frequency applied to the ring electrode; and (4) by selecting a point on the stability diagram other than the normal point ($q_z = 0.908$, $a_z = 0$) to cross from stability to instability. Previously, only a few attempts had been made to increase the m/z limit beyond 650 Da per charge. Stafford and Syka[37] successfully increased the m/z limit by method (2), changing the trap dimensions such that $r_0 = z_0$. In this configuration, a mass limit of approximately 850 Da was achieved, yielding a m/z increase of 1.3X compared with that of the commercial system. Method (4) for m/z range extension has been reported by Todd et al.,[38] who introduced the concept of performing mass analysis scans at working points on an instability boundary other than that ($\beta_z = 1$) on which the point $a_z = 0$, $q_z = 0.908$ is located.

The mass range extension was achieved in a reverse scanning technique by using a positive DC voltage and ramping the RF amplitude downward while maintaining a constant RF/DC ratio. By ejecting ions at the (β_z = 0 boundary line, a theoretical m/z limit of 2185 Da (3.4X increase) is possible at q_z = 0.27. Using this method, ions at m/z 1466 derived from tris(perfluorononyl)-s-triazine have been ejected mass selectively from the ion trap and detected externally.[38] Three different techniques aimed at extending the m/z range of the ion trap have been explored in our laboratories: reduction of the dimensions (both r_0 and z_0) of the theoretical trap by $\frac{1}{2}, \frac{1}{3}$, and $\frac{1}{4}$ as suggested by method (2); reduction of the RF frequency applied to the ring electrode as suggested by method (3); and application of a supplementary AC voltage across the end-cap electrodes to resonantly eject ions from the trap. Resonant ejection was also combined with frequency reduction to achieve the detection of the highest mass singly charged ions (44,560 Da per charge) to date.[27]

B. Reduced Size Electrodes

The electrode design used for the reduced size electrodes conformed to theoretical calculation for the surfaces and for the spacings between the electrodes.[39] The end-cap electrodes have surfaces corresponding to hyperboloids of two sheets:

$$\frac{r^2}{r_0^2} - \frac{z^2}{z_0^2} = -1 \qquad (4)$$

The ring electrode has a surface corresponding to a hyperboloid of one sheet:

$$\frac{r^2}{r_0^2} - \frac{z^2}{z_0^2} = 1 \qquad (5)$$

The spacing between each end-cap electrode and the ring electrode followed the equation:

$$r_0^2 = 2z_0^2 \qquad (6)$$

Two sets of small electrodes, half- and quarter-sized, were characterized using Cs^+ SIMS. Figure 3a shows the mass spectrum of $(CsI)_nCs^+$ using the half-sized trap. The complete sequence of clusters $(CsI)_nCs^+$ is observed and the peak at m/z 2473, due to $Cs_{10}I_9^+$, is near the theoretical limit of the device. Resolution is approximately 1500 ($m/\Delta m$, where Δm is FWHM) at 2500 Da. Note, however, that artifact signals at low intensity occur. It is believed that these are due to imperfections in the RF field caused by the relatively large holes ($\frac{1}{16}$ in. diameter) and the use of the ideal rather than the stretched geometry. Measures to reduce these field distortions were not attempted. CsI cluster ions were reanalyzed with the quarter-sized electrodes (theoretical m/z limit is 10.4 kDa per charge). Figure 3b illustrates the mass spectrum in the range of 6 to 10 kDa per charge. Note again the high abundances of ions observed out to the mass limit along with the presence of enhanced artifact peaks among the sequence of cluster ions. The resolution for this experiment was measured to be approximately 1900 for $Cs_{38}I_{37}^+$ at m/z 9753. It is postulated that improvements in the quality of the trapping fields would lead to enhanced resolution.

A reduction in the RF voltage ramp rate to produce a reduction in the m/z scan rate should further improve spectral resolution. The RF ramp rate of the commercial system provided for a scan rate is 5555 Da s^{-1}. Upon increasing the mass range by a factor of 4 or 16 (by reducing the size of the electrodes by $\frac{1}{2}$ and $\frac{1}{4}$, respectively) the rate is increased to 22,220 Da s^{-1} and 88,880 Da s^{-1}, respectively. These scan rates have been found to be too fast for obtaining good spectral resolution. Reduction of the scan rate to the normal 5555 Da s^{-1} by simply altering the instrument software via the introduction of an additional time delay after every RF DAC step during the scan ramp enhances the resolution obtained using the quarter-sized trap by approximately a factor of 2. This method of reducing the scan rate is different from slowing the scan by attenuating the voltage derived from the mass analysis DAC (see high resolution section below), and is limited in performance due to the fact that when delays are protracted, a stairstep ramp is applied to the ring electrode instead of a linear ramp.

C. Reduction of the RF Drive Frequency

Reductions in the frequency of the RF applied to the ring electrode from the standard 1.1 to 0.92 MHz and 0.60 MHz were also studied. Note that frequency reduction was accompanied by changes in the RF amplitude that could be obtained; therefore, the theoretical mass limits are not predicted simply by the square of the change of the frequency, as indicated by Equation 3. Figure 4 illustrates a mass spectrum of RbI desorbed with 7 kV Cs$^+$ projectiles, recorded at a reduced drive frequency of 0.6 MHz. The observed m/z limit is approximately 2050 Da per charge (3.15X mass range extension). As indicated in this mass spectrum, RbI ions form stable cluster ions much like CsI. The rate of the mass scan in this mass spectrum is 16,666 Da s^{-1} and the resolution is approximately 450 at m/z 933 ($Rb_5I_4^+$) and twice this at m/z 1781 ($Rb_9I_8^+$). Reduction of the scan rate by introduction of a variable delay (by modification of the ITMS™ software) after every DAC step in the mass analysis RF scan was performed, but resulted in only a modest improvement in resolution. At low m/z ratios with the scan rate slowed using this method, resolution was still

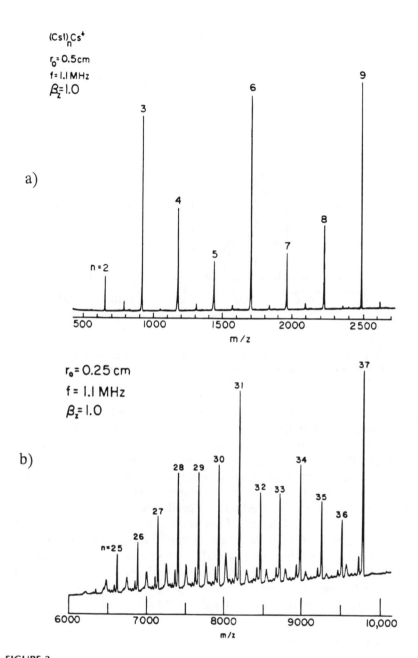

FIGURE 3

Mass spectrum of $(CsI)_nCs^+$ obtained using Cs^+ SIMS ionization and ion injection into (a) the half-sized ion trap ($r_0 = 5$ mm); (b) the quarter-sized ion trap ($r_0 = 2.5$ mm). (Adapted from Kaiser, R.E. et al., *Int. J. Mass Spectrom. Ion Proc.*, 106, 79, 1991.)

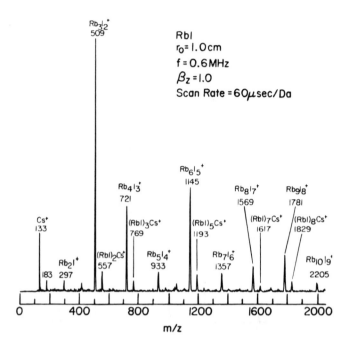

FIGURE 4

Mass spectrum of RbI. The RF drive frequency of the ion trap was reduced from 1.1 to 0.6 MHz. The commercial set of electrodes ($r_0 = 1$ cm) was used. (Adapted from Kaiser, R.E. et al., *Int. J. Mass Spectrom. Ion Proc.*, 106, 79, 1991.)

worse than that observed when the trap is operated without modification. These data demonstrate that it is possible to extend the mass range of the ion trap as predicted by Equation 3 and indicate that resolution decreases at the lower drive frequency even when the effective scan rate is reduced (by using appropriate time delays in the stepping voltage, V) to a rate similar to that of the standard system. However, when the scan rate is attenuated using a method that produces a smooth ramp (such as the offset DAC method described below), much of the lost resolution can be regained. Reduction in the fundamental RF drive frequency, although electronically somewhat complicated, is an effective way to extend the mass range. In fact, when utilized together with resonance ejection at appropriate frequencies (also discussed below), it can yield a higher mass range with *increased* resolution at a given mass scan rate.[40]

D. Resonance Ejection

The final method evaluated for mass range extension employs the mass-selective instability scan and uses a supplementary AC voltage applied to the end-cap electrodes to eject ions from the trap during this scan.

This method utilizes the long-established method of resonance excitation.[14] The difference between the resonance ejection method and resonance excitation as applied in MS/MS experiments lies in the timing and in the intensity of the applied supplementary AC field. In tandem experiments, the supplementary AC voltage amplitude is typically below 1.5 V_{0-p} and is applied prior to mass analysis with the RF at a fixed amplitude. For resonance ejection experiments, the AC amplitude is typically > 3.0 V_{0-p} so that ions, when brought into resonance with the supplementary field, are excited to such an extent that they are ejected from the ion trap. When the resonance ejection experiment is performed at a frequency close to half the RF drive frequency (β_z is close to 1) and q_z is scanned, enhanced resolution and sensitivity are achieved when compared to data obtained without resonance ejection. This technique, not intended to extend the m/z range, is called axial modulation.[41] Spectral peaks, broadened to a width of about 5 Da by space charge, can be reduced in width to about 1 u with this technique, resulting in improved resolution and sensitivity.

By reducing the frequency applied to the end-cap electrodes, ions can be ejected from the trap even though their stability coordinates lie within the stability diagram. The effect of the supplementary AC field on the stability diagram can be visualized as a line of instability within the stable region. The intersection of this iso-β_z line (or iso-β_r line) with the mass-selective instability scan line, which is the q_z-axis when the instrument is operated in the "RF-only" mode, creates a hole in the stability diagram (Figure 5) which is denoted by q_{eject} (the value of q_z at which ejection occurs).

The value of q_{eject} defines the factor by which the mass range is extended when resonance ejection is used. Therefore, it determines the corresponding β_{eject} value at which a resonance instability must be generated. As shown by Equation 3, mass is inversely proportional to q_z, but it is a complex function of β_z. The q_{eject} value necessary for a desired mass range extension $m/z_{limit-new}$ can be calculated from the following relationship:

$$q_{eject-new} = q_{eject-old} (m/z_{limit-old})/(m/z_{limit-new}) \qquad (7)$$

where $q_{eject} = 0.908$ and $m/z_{limit-old} = 650$ Da per charge for the standard Finnigan MAT ion trap. The conversion of q_z to β_z cannot be expressed in closed form; however, numerical procedures to accurately calculate β_z from q_z are relatively straightforward.[13] Once β_{eject} is known, the frequency needed to ejected ions for a specified m/z range extension factor can be calculated from the following relationship:

$$\omega_z = \beta_z \Omega_0/2 \qquad (8)$$

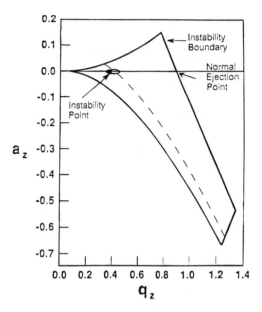

FIGURE 5
Stability diagram illustrating resonance ejection at the intersection of the scan line and the iso-β line. (Adapted from Cooks, R.G. et al., *Chem. Eng. News*, 69, 26, 1991.)

where ω_z is the fundamental axial secular frequency. For example, if the mass range is extended by a factor of 5.00, q_{eject} will be 0.182, and the appropriate β_z value for ejection will be 0.127, as obtained from the numerical procedure.[42] The β_z value corresponds to a secular frequency of 69.9 kHz (or a angular frequency of $2\pi \times 69.9$ krad s^{-1} for $\Omega_0 = 2\pi \times 1.1$ MHz), which is applied to the end-cap electrodes during the mass-instability scan.

In an attempt to determine the ultimate mass limit of the unmodified commercial ion trap, experiments were carried out on cluster ions generated by bombarding solid CsI with 7 keV Cs$^+$ ions. The resonance ejection frequency was reduced systematically, the lowest value used being 4600 Hz. When the applied frequency is 35.2 kHz, ejection occurs at $q_z = 0.091$, which corresponds to $\beta_z = 0.063$ and $(m/z)_{max} = 6500$ (10 times the m/z range of the trap when the mass-selective scan is used without resonant ejection). Figure 6 illustrates the mass spectrum of CsI showing clusters up to $n = 22$. The resolution achieved at m/z 3510, without any modification of the scanning speed, was approximately 3000 FWHM. In principle, the m/z range can be extended indefinitely by choosing an extremely low β_{eject} value without further modifications to the ion trap as shown in Figure 7. However, an increase in excess of 60-fold in the m/z range by resonant ejection is precluded by limitations of the electronics, which generate the supplementary AC voltage, rather than by any fundamental limitation of the technique. For a 60-fold increase, the m/z range will be approximately 39 kDa per charge.

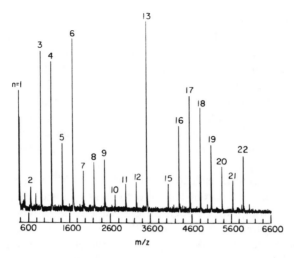

FIGURE 6

Mass spectrum of CsI clusters acquired using a resonance frequency of 35.2 kHz (q_{eject} = 0.0904, β_z = 0.064). Cluster sizes of $(CsI)_nCs^+$ are indicated by values of n. (Adapted from Kaiser, R.E. et al., *Int. J. Mass Spectrom. Ion Proc.*, 106, 79, 1991.)

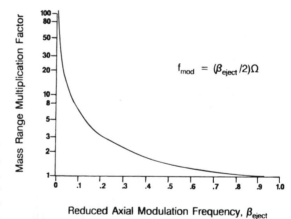

FIGURE 7

Illustration of how the mass range is increased using resonance ejection as the value of β_z for ejection is decreased.

$$f_{mod} = (\beta_{eject}/2)\Omega$$

Reduced Axial Modulation Frequency, β_{eject}

Because no loss in performance was observed when the main RF frequency was decreased to 0.92 MHz, resonance ejection experiments aimed at increasing the m/z range further were performed at this RF frequency. This alteration gave an additional 1.7-fold increase to the mass range. A resonant ejection frequency of 4600 Hz (β_{eject} = 0.01) was then applied to the end-caps, giving an expected m/z limit of approximately 72 kDa (111-fold increase in m/z). Figure 8 illustrates a region of the CsI mass spectrum from 22 to 46 kDa. The base peak is $(CsI)_{122}Cs^+$ (n = 122) m/z 31,853.

Recently, high mass experiments were performed at Finnigan MAT on an ion trap equipped with a somewhat tunable RF drive frequency.[23] The highest mass range extension was achieved using a combination of a reduced fundamental RF frequency of 0.550000 MHz and a resonance ejection frequency of 8832 Hz ($\beta_{eject} = 0.0321$, $q_{eject} = 0.0454$). The resulting mass range was extended to 52 kDa and singly charged egg albumin ions (mol wt 43.3 kDa), generated by laser desorption, were injected into the trap and mass analyzed (Figure 9). Although methods exist to improve mass resolution (see below), no resolution enhancement techniques were utilized in this experiment.

Resonance ejection also can be utilized in MS/MS experiments to isolate selected high m/z parent ions and to mass analyze fragment ions. A typical timing diagram for such an experiment is shown in Figure 10. After the ionization period, a reverse-then-forward RF scan is performed while the supplementary AC is applied to the end-cap electrodes.

This isolation technique is used because high m/z ions cannot be isolated at the upper apex of the stability diagram using conventional RF/DC isolation. In the reverse portion of the isolation scan, ions of m/z greater than the selected precursor ion(s) are ejected from the trap along with

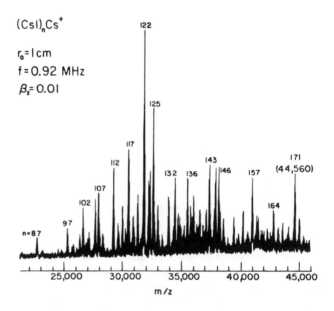

FIGURE 8

Mass spectrum of CsI acquired using a resonance frequency of 4600 Hz ($q_{eject} = 0.02$, $\beta_z = 0.014$). The drive frequency was reduced from 1.1 to 0.92 MHz to achieve an additional 1.7-fold increase in mass. Cluster sizes are as indicated in Figure 6. (Adapted from Kaiser, R.E. et al., *Int. J. Mass Spectrom. Ion Proc.*, 106, 79, 1991.)

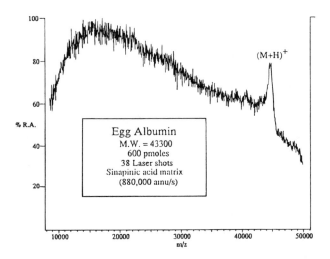

FIGURE 9
Matrix-assisted laser desorption mass spectrum of egg albumin.

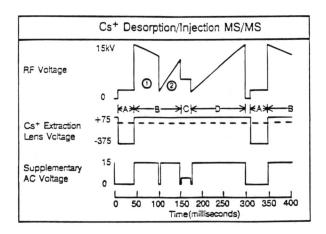

FIGURE 10
Reverse-then-forward scan function for precursor ion isolation in a resonance ejection MS/MS experiment. (Adapted from Kaiser, R.E. et al., *Rapid Communication in Mass Spectrometry*, vol. 3, Heyden and Sons, 1989, 225.)

ions (>m/z 650) that are ejected from the trap due to crossing the instability boundary at the point $q_z = 0.908$, $a_z = 0$. The supplementary AC is then turned off to allow the precursor ion(s) to remain trapped while the RF amplitude is lowered to perform the forward scan. The supplementary AC is then turned on and the RF amplitude is increased to eject all of the remaining low mass ions. Figure 11a to c shows the signal produced at three different times, (a) prior to isolation, (b) after the reverse

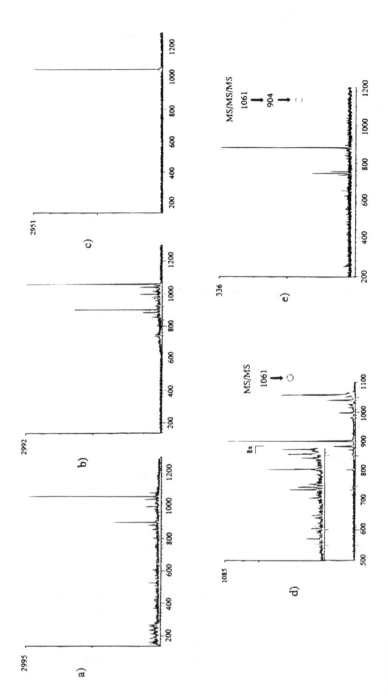

FIGURE 11

Parent ion isolation of bradykinin (mol wt 1060) using reverse-then-forward isolation (a to c); (d) MS/MS; (e) MS[3]. (Adapted from Cooks, R.G. and Cox, K.A., Biological Mass Spectrometry: *Present and Future*, Matsuo, T. et al., Eds., John Wiley and Sons, 1994, 179.)

RF scan, and (c) after the forward RF scan, when the nonapeptide bradykinin (average mol wt 1060) is analyzed. The signal-to-noise (S/N) ratio is improved greatly due to the removal of the matrix ion background along with fragment ions created during ionization.

Once the precursor is isolated, the RF amplitude is set at an appropriate q_z value to allow resonance excitation. A supplementary AC voltage is applied at amplitudes low enough to excite resonantly, rather than eject, the ions so that they gain kinetic energy and collide with the helium buffer gas to produce fragment ions. Typically, higher mass precursors require longer excitation periods along with higher amplitudes of the supplementary AC to induce fragmentation when compared to typical excitation conditions for low mass precursor ions. Efficiencies for high mass MS/MS experiments are typically lower than those observed for low mass organic species. Tandem experiments on higher mass ions (> 5000 Da per charge) are limited because the maximum RF amplitude allows resonant excitation only at low q_z values (< 0.15), and dissociation efficiencies are reduced dramatically. The MS/MS spectrum of the isolated $(M + H)^+$ ion is shown in Figure 11d. The m/z range was extended to 1300 Da per charge by applying an AC frequency of 184.6 kHz to the end-cap electrodes to eject the ions resonantly. Multistage MS of high mass ions also can be performed. The previous experiment was continued by the addition of an isolation period (by another reverse-then-forward scan) and a resonance excitation period (Figure 11e) to dissociate a fragment ion in an MS^3 experiment.

E. Future

Although several methods of extending the m/z range of the ion trap exist, clearly, the single most effective method demonstrated to date is resonance ejection. This technique can be implemented easily on any ITMS with only minor modifications to the electronics. Also, mass range extension by resonance ejection allows the trap to be used to obtain high resolution mass spectra (see below). The ultimate ion trap for high mass work has yet to be determined. Such a system could consist of electrodes reduced in size and operated with a modest reduction in frequency using the mass-selective instability scan mode combined with resonance ejection.

IV. HIGH RESOLUTION[27-32]

A. Introduction

A significant improvement in mass resolution is achieved when ion traps are operated in the presence of helium.[13] In the mass-selective instability mode,[13] the Finnigan MAT ITMS™ has unit resolution for ions with m/z ratios <650. Axial modulation,[41] a technique in which a low amplitude AC voltage is applied to the end-cap electrodes at a frequency slightly less than half of the RF drive frequency, provides a more definite instability point for ions detected using the mass-selective instability scan, and results in mass resolution up to 2500 ($m/\Delta m$) of ions with masses ≤650 Da per charge. Further resolution enhancement in the ion trap can be achieved by a reduction in the scanning rate of either RF voltage[27-33,43] or AC frequency[44] (which can be scanned) when the mass-selective instability mode of operation is used. The most common method of resolution enhancement is through the reduction of the RF scanning rate. The offset-DAC technique, which is used to slow the scan rate, is discussed in detail in this section.

B. Offset-DAC

The mass analysis scanning rate is normally fixed at 180 μs Da^{-1} (5555 Da s^{-1}) in the commercial ion trap operated in the mass-selective instability mode with axial modulation. These operating conditions result in a resolution (FWHM) of approximately $3 \times m$, where m is mass and singly charged ions are assumed.[45] When operating with the standard (650 Da) mass range and m/z scan rate, the time interval between successive output steps of the 12-bit DAC controlling the RF voltage amplitude is 28 μs. This time is sufficiently short that the RF control system is unable to track individual DAC output steps. Hence, the RF amplitude appears, to a good approximation, as a continuous linear ramp. When operating with an extended mass range, the time interval between RF DAC steps can be extended in order to produce m/z scan rates of 5555 Da s^{-1}. However, in this case, the amplitude of the RF drive voltage tracks the stair step waveform output by the RF DAC. Furthermore, these steps are potentially quite coarse in terms of the incremental change of the mass at the threshold of trajectory instability with each step of the RF amplitude. For example, when the mass range is extended to 6500 Da, a 10-fold extension from the standard mass range, a 1.6-Da change occurs in the mass of ions per RF DAC output increment. This coarseness in the change in the mass at the threshold of instability per DAC output step scales proportionally with the degree of mass range extension, no matter what method of mass range extension is used.[27] Consequently, this method of reducing the scan rate is severely limited.

The design of the commercial ion trap is also such that the time interval between analog-to-digital converter (ADC) samples of the ion current signal is always the same as the interval between RF DAC increments. Therefore, the density of ADC samples per unit mass varies inversely with the factor by which the mass range is extended. The ADC sample density of the system, when operated with its standard mass range, is only 6.3 ADC samples per Dalton. When the mass range is extended beyond 4100 Da, the sample density drops to <1 ADC sample per Dalton. Furthermore, information theory and experience dictate that it requires as a minimum about six samples across a single mass peak in order to represent its peak shape with reasonable accuracy. This requirement means that without special modification, the instrument will have great difficulty in representing accurately any mass peaks that have a FWHM resolution in excess of about 1800, no matter what the mass range of the instrument. This situation correlates well with the upper limits of resolution observed in the previously described experiments.[27]

In an attempt to circumvent these limitations without undertaking a complete redesign and replacement of the existing RF control and data acquisition system, an RF control scheme utilizing a supplementary DAC was implemented.[27] The basic idea of the scheme was that if the scaling between the RF DAC output and the actual RF voltage amplitude applied to the ion trap ring electrode was reduced in inverse proportion to the factor by which the mass range of the instrument was being extended, the number of DAC increments and ADC samples per mass unit (and per unit time) during the mass analysis RF voltage ramp would remain unchanged. However, doing this alone would not solve the entire problem, as the mass range over which mass spectral data could be acquired would remain unchanged at 650 Da, even though the intrinsic mass range of the instrument would be extended. This limitation was overcome by connecting the output of a supplementary DAC into the RF amplitude control circuitry so that the amplitude of the RF would be determined by the scaled sum of the RF DAC and supplementary DAC outputs. This arrangement enables the full mass range of the instrument to be accessed while maintaining appropriate RF DAC scaling and ADC sampling density. The drawback of this scheme is that a limited portion of the entire mass range may be scanned in any one experiment. The size of this mass window is determined by the attenuation factor of the RF DAC and the mass range extension factor.

A schematic illustration of this offset-DAC method is shown in Figure 12. The normal RF ramp is governed by a 0 to 10 V RF DAC output applied to the ring electrode. This ramp can be attenuated by a voltage divider at any time in the scan sequence through a TTL control pulse. The time taken to traverse the voltage ramp is the same; however, a much narrower voltage range is scanned. Any region in the normal 0 to 10 V scan (any mass range, either m/z 0 to 650 in the normal operation of the ion trap, or extended mass ranges when resonance ejection is employed) can be observed by supplementing the attenuated ramp with an appropriate offset voltage. The final RF control signal (attenuated DAC voltage plus offset voltage) is then passed through an RF generator to achieve the final RF voltages necessary for mass analysis. A detailed circuit schematic is described elsewhere.[27]

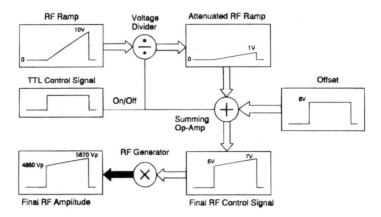

FIGURE 12
High resolution mass analysis via offset-DAC method to reduce RF ramp rate.

Using the offset-DCA circuit, experimentally actual resolution enhancements of the MH$^+$ ion of substance P (mol wt 1348) are illustrated graphically in Figure 13.[32] The resolution is measured as mass over the FWHM of the individual peaks in the isotopic cluster.

C. Mass Spectra

1. Applications

In single stage (MS) experiments, resolutions as high as 350,000 ($m/\Delta m$) have been observed in the case of peptides (mol wt 1349),[46] and in the case of the CsI salt cluster (mol wt 3510), resolutions in excess of 10^6 have been recorded.[31] The highest mass resolution demonstrated exceeds 10^7 for the ion at m/z 614 generated from perfluorotributylamine (FC-43) using a more highly modified offset DAC circuit.[33] In practice, resolution is enhanced only to the point of achieving better than unit resolution of singly charged ions in MS and MS/MS spectra for structure elucidation.[32] Higher resolutions are employed for the separation of the isotopic forms of multiply charged ions in order to identify the charge state observed in electrospray spectra.[28,30]

One application of increased resolution can be seen in the detection of the reduction of a single disulfide bond during Cs$^+$ desorption ionization shown in Figure 14.[32] A disadvantage of ionization from a liquid matrix is that reduction can occur readily during ion bombardment,[47] and it is, therefore, not surprising that the disulfide bond in somatostatin is reduced to some extent while its mass spectrum is recorded (Figure 14a). The all-^{12}C-form of MH$^+$ is shifted from the expected value of 1637 Da to a new value of 1639 Da. Complete reduction can be achieved by deliberately adding a reducing agent, dithiothreitol, to the matrix (Figure 14b).

Resolution as a Function of Scan Speed
(M+H)+ of Substance P

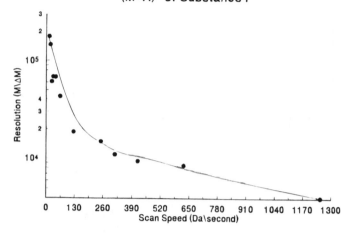

FIGURE 13

Experimentally determined resolution enhancements of the MH· ion of substance P obtained by slowing the data acquisition scan speed. (Adapted from Cox, K.A. et al., *Biological
Mass Spectrometry*, vol. 21, John Wiley and Sons, 1992, 226.)

Note that the convention adopted in presenting high resolution results is
to use the mass of the known $^{12}C^1H^{14}N^{16}O$ isotopic species and to round
down in order to count the spectrum in integral steps. This convention
becomes extremely important when calibration procedures are employed
for accurate mass assignments (discussed later).

When dealing with multiply charged ions such as those produced by
electrospray ionization, it is necessary to know the charge state of an ion
in order to determine mass. When peaks arise from a series of consecutive charge states, which is often the case for electrospray mass spectra,
then the number of charges can be calculated. However, when no associated charge states are known, as in the case of MS/MS spectra of multiply charged precursor ions, the charge on a given ion species and, consequently, the mass remain unknown. This problem becomes particularly
extremely important in structure elucidation, such as peptide sequencing, where MS/MS of a multiply charged parent ion is often used to obtain sequence information. This problem is discussed in more detail later.
One solution to this problem is to use high resolution to determine the
spacing between the isotopic peaks in an ion of a single charge state.

The electrospray ionization mass spectrum of horse angiotensin I is
shown in Figure 15.[28] This spectrum was obtained by extending the mass
range by a factor of 3 and using resonance ejection of 119,936 Hz ($q_{z\text{-eject}}$
= 0.30), resulting in a scan speed of 11,650 Da s^{-1}. The singly, doubly, triply,

a
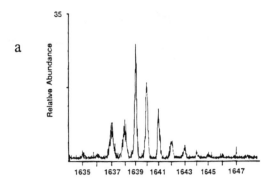

b
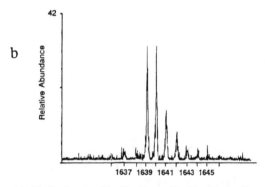

Ala-Gly-Cys-Lys-Asn-Phe-Phe-Trp-Lys-Thr-Phe-Thr-Ser-Cys

FIGURE 14
High resolution mass spectrum of the MH⁺ region of somatostatin (monoisotopic mol wt 1636). Unit resolution of the isotopes shows (a) partial reduction of the disulfide bond and (b) complete reduction upon the addition of dithiothreitol to the sample. (Adapted from Cox, K.A. et al., *Biological Mass Spectrometry*, vol. 21, John Wiley and Sons, 1992, 226.)

and quadruply protonated ions are indicated. Scanning more slowly across the multiply charged ions using a scan rate of 278 Da s⁻¹ and a resonance ejection frequency of 464,371 Hz provides isotopic separation of these species as displayed in the inset windows. This scan speed yields a peak width (approximately 0.087 Da) FWHM that readily allows identification of the charge state for these molecular ions.

In addition to resolution of isotopic ions, isobaric separation[43,48] also has been achieved using the offset DAC technique. Figure 16 shows the separation of low mass isobars, which differ in mass by 0.087 Da, produced by Xe⁺ and $C_3F_5^+$). The scan rate was slowed 11.1 Da s⁻¹ to obtain this resolution. Figure 17 shows a portion of a mass spectrum obtained of co-injected CsF and substance P when the scan rate was slowed to 130.5 Da s⁻¹; the co-injection procedure is discussed in more detail below.

2. Simulations

Computer simulations have proven extremely useful in understanding the complex behavior of ions in the ion trap. Two simulation pro-

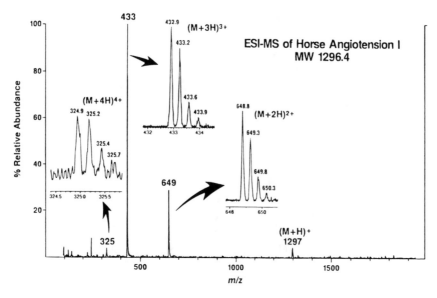

FIGURE 15

Electrospray mass spectrum of horse angiotensin I at 16,650 Da s⁻¹. Insets show multiply protonated molecules scanned using 278 Da s⁻¹ scan rate.

grams have been used at Purdue University; a small scale simulation, ion trap simulation (ITSIM),[49] which can be used on a personal computer and monitors the behavior of <1000 ions, and a large-scale simulation, numerical quadrupole simulation (NQS),[50] which is run on a mainframe computer and has the ability to monitor larger numbers of ions. Figure 18 compares a simulated high resolution mass spectrum of the molecular

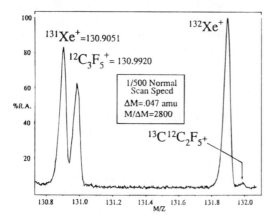

FIGURE 16

Portion of an electron ionization mass spectrum of a mixture of Xe and FC-43. The spectrum was scanned at a rate of 11.1 Da s⁻¹.

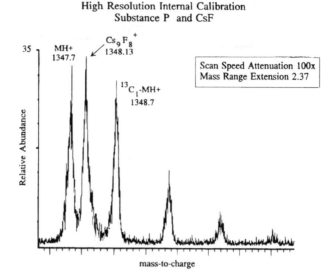

FIGURE 17

Portion of a high resolution mass spectrum showing the separation of $Cs_9F_8^+$ and the all-^{12}C and $^{13}C_1$ protonated molecular ions of substance P.

ion region of substance P (a) with the experimentally determined spectrum (b). In each case, the scan rate was reduced by a factor of 10, giving a scan rate of 1250 Da s⁻¹ with the m/z range extended to 1463 Da per charge. The ions are identified in terms of their RF voltage upon ejection, which is directly related to mass. The slight shift in voltage between the two spectra corresponds to approximately 1 Da, which is within the experimental error of the probe used to measure the voltages. It is remarkable that good agreement in peak shapes and resolution is observed between the two spectra even though no buffer gas or space-charge interactions were incorporated into the simulation.

D. Disadvantages

A drawback to the high resolution experiments can be seen in the high resolution spectrum of the molecular ion region (MH⁺) region of gramicidin S, shown in Figure 17.[32] In this case the offset-DAC circuit was employed to slow the acquisition scan rate from 5555 to 131 Da s⁻¹. Note the nonuniform spacing between the consecutive isotopes. The space observed between the $^{13}C_2$- and $^{13}C_3$- isotopes is clearly larger than the spacing between the all-^{12}C and $^{13}C_1$ isotopes. The nature of these shifts is quite complex and not completely understood, although it is known that ion motion is highly dependent on the other ions present in the trap during

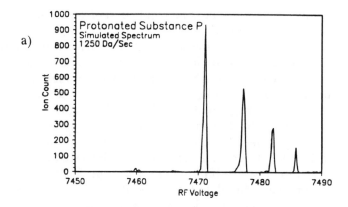

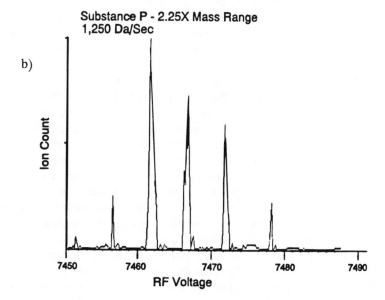

FIGURE 18

Comparison of (a) simulated and (b) experimental high resolution mass spectra of the MH⁺ region of substance P.

ejection. Clearly, these shifts will affect the mass measurement accuracy that can be obtained in these experiments.

Another difficulty observed in these experiments is that at very low scan speeds, a slight shifting of the peak position occurs with each scan, and most likely within a single scan. Consequently, when many scans are averaged, the observed peak width is broadened. This drift is at least partly due to RF instability, which can be observed only at these low scan rates and narrow peak widths. It is possible that with appropriate hardware modifications this problem can be rectified, thereby allowing even greater resolution to be achieved.[28]

The S/N ratio decreases as the scan rate is decreased, prohibiting high resolution measurements on low abundance species. Although no ions are lost when the scan rate is reduced, the detector measures a reduced ion current. In other words, the ion signal is spread out in time, which reduces the signal height without affecting the height of the noise. Appropriate filtering of the data so that the bandwidth of the detection system matches the achieved resolution for a given scan rate can gain back some of the S/N. There have been cases, such as studies with CsI clusters, in which the peak width in time (or ion current) remains constant as the scan rate is reduced. In these special cases, no loss in S/N is observed. Also, the tolerance for space charging decreases as the scan rate is reduced. This is particularly evident when multiply charged ions are mass analyzed. Thus, in many cases, the number of ions in the trap must be reduced dramatically in order to achieve high resolution; under these circumstances, a true loss of signal occurs and the S/N suffers. For best results, only the ions of interest and appropriate mass standards should be stored in the trap for high resolution experiments.

One other limitation of the high resolution technique is that the resolution that can be attained for a particular ion is dependent on its thermochemical stability. If an ion fragments readily, as the ion gains kinetic energy during the slow resonance ejection scan, fragmentation will occur and the product ion will be ejected, thus leading to tailing on the low mass side of the peak, which degrades the resolution. Consequently, very high resolutions are best obtained with fairly stable ionic species.

A detailed theoretical discussion of the enhanced resolution experiment is provided elsewhere.[51] This work examines the relationship between mass and resonance excitation frequency line widths, allowing the mass resolution to be evaluated as a function of fundamental ion trap operating parameters. The results from the theoretical study correlate reasonably well with experimental results; however, recent results by Londry et al.[33] show experimental resolutions in excess of ca. 10^2 beyond those predicted by this model. More work is needed to understand all the parameters that affect mass resolution.

V. HIGH RESOLUTION IN TANDEM (MS/MS) EXPERIMENTS

A. Isolation

The recently recognized capability for obtaining ion trap spectra at higher resolution through the use of slow scans has the potential to allow individual isotopic forms of the protonated molecule to be employed in

MS/MS experiments. The same capability should also allow product spectra to be recorded at unit mass resolution. Note, however, that it may not be necessary to isolate a particular isotopic form of MH⁺ in order to obtain the product tandem mass spectrum exclusively of that isotopic form. Because each isotope has a different secular frequency, characteristic of its m/z ratio, it is possible to excite resonantly and independently one isotopic ion in the MH⁺ cluster. In this case, all observed products arise from just the selected (but not isolated) isotopomer. In practice, however, it is often extremely difficult to selectively excite a single isotope due to the high amplitude of the excitation voltage required, especially for high mass ions, and the close secular frequencies of the neighboring isotopes. Thus, isolation of the isotope is often required for unambiguous formation of products from a single isotopic parent ion.

A high resolution isolation experiment is illustrated in Figure 19 for the triply charged electrosprayed ion of renin substrate at m/z 587.[30] Isolation can be accomplished using either a reverse-then-forward[32] or

$(M+3H)^{3+}$ Renin Substrate

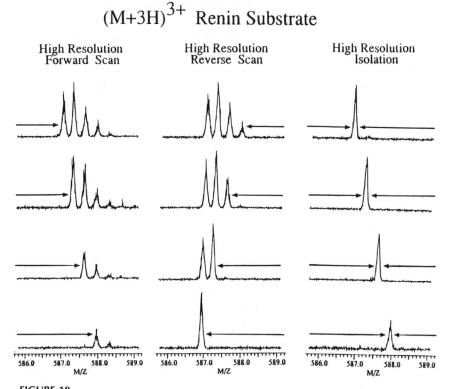

FIGURE 19
High resolution forward scan, reverse scan, and ion isolation of the $(M + 3H)^{3+}$ ion of renin substrate.

forward-then-reverse[30] RF scan. In order to achieve high resolution isolation, resonance ejection must be used. The high resolution isolation procedure employs two forward-then-reverse isolation sequences starting with a low resolution isolation using normal RF scans to select a small mass window, and then another isolation utilizing the slower RF scan. The spectrum in the top left corner of Figure 19 shows the entire isotopic distribution resolved using a normal high resolution scan performed at $\frac{1}{100}$ the normal RF scan rate at an auxiliary field frequency of 360 kHz and an amplitude of 3.2 V_{0-p}. Each spectrum in Figure 19 consists of averaging 120 scans in approximately 40 s. The total sample consumption was 4 pmol, although only 24 fmol was actually ionized and injected into the analyzer. The isotopes are spaced by m/z 0.33 with a peak width of m/z 0.09 and a resolution of approximately 6500 FWHM. Shown in the left-hand column is the effect of the high resolution forward scan which also uses an auxiliary field frequency of 360 kHz and an amplitude of 3.2 V_{0-p}, indicating that lower mass isotopes can be eliminated consecutively with no apparent loss of the higher mass isotopes. Similar data are obtained (center) for high resolution reverse scans. The results of the combination of the high resolution forward and reverse scans are shown at right in Figure 19, and demonstrates the isolation of individual isotopes. Note that the vertical scales on each spectrum are not exactly the same due to the normalization routines of the data system; however, it is clear that the S/N does not change significantly in any of the spectra.

B. Applications

An examination of the fragmentation of substance P with irradiation of the entire MH^+ cluster (m/z 1347 to 1351) gives the partial spectrum shown in Figure 20(a).[32] This spectrum shows the NH_3 and H_2O loss occurring from all the isotopic forms of the protonated molecule. When the ^{12}C isotope of the protonated molecule is isolated in the trap and dissociated, the two peaks shown in Figure 20(b) result. A similar spectrum, appropriately shifted in mass, is recorded when the $^{13}C_1$- isotopic form is isolated and its product spectrum is recorded (Figure 20(c)). This type of high resolution MS/MS experiment assists in the elucidation of fragmentation mechanisms. For example, the m/z 1330 product (mass measurement accuracy ±0.1%) observed in a low resolution tandem mass spectrum of substance P could be due to H_2O loss, NH_3 loss, or a combination of the two. High resolution of the product region reveals it to be due to a mixture of the two processes. Dissociation of a single parent isotope confirms that this fragment arises by both H_2O and NH_3 losses. Because the product ions resulting from the dissociation of the $^{13}C_1$ isotope overlap with those resulting from the ^{12}C isotope (at m/z 1330), isolation of an

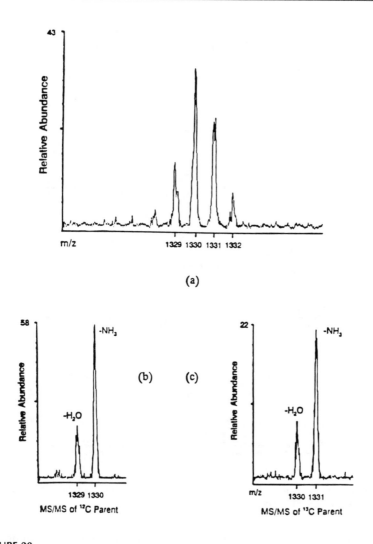

FIGURE 20

MS/MS experiment on substance P. (a) Resolved fragment (loss of H_2O and NH_3) from the unresolved MH^+ parent; (b) loss of H_2O and NH_3 from the ^{12}C isotope; and (c) loss of H_2O and NH_3 from the ^{13}C isotope. (Adapted from Cox, K.A. et al., *Biological Mass Spectrometry*, vol. 21, John Wiley and Sons. 1992, 226.)

individual isotope prior to dissociation is necessary for unambiguous determination of fragmentation behavior.

High resolution MS/MS experiments performed on multiply charged ions can be used to determine unknown fragment ion charge states. Figure 21 illustrates the utility of this type of experiment on the quadruply charged molecular ion, $(M + 4H)^{4+}$ of renin substrate at m/z 440.[28] By activating the

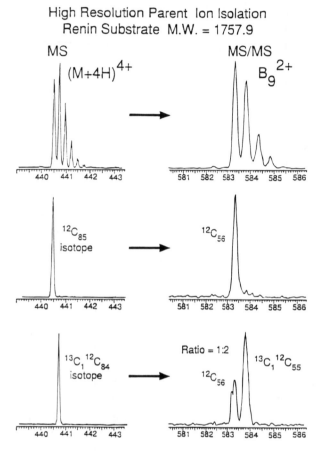

FIGURE 21
High resolution isolation of multiply charged isotopes and high resolution of the fragment
ions for renin substrate (mol wt 1757.9).

entire cluster, the multiply charged fragment ions will also contain isotopes that can be resolved, the spacing of which allows the determination of the charge state (as shown for the b_9^{24+} fragment ion). Again, the
MS/MS spectrum can be simplified by isolating and dissociating a particular isotope of the precursor ion. Dissociation of the ^{12}C isotope reveals
the fragment ion to be monoisotopic, while dissociation of the $^{13}C_1$ isotope produces a ratio of the all-^{12}C and the $^{13}C_1$ species indicative of the
number of carbons present in the fragment ion.

C. Future

At this point in ion trap development, these types of experiments are
by no means routine. As mentioned previously, the behavior of the ions

in the trap is dependent on experimental conditions involving the presence of other ions and neutral molecules in the trap. Also, instabilities in the existing electronics contribute to peak broadening, degrading the expected resolution. However, these hurdles are typical of any new technique in the early stages of development; experimental studies are currently underway to understand and to eliminate these drawbacks. There can be no doubt of the power of this method, especially in the areas of biochemistry and biotechnology. High resolution information can be gained on molecular ions. Isolation of single isotopic species can be achieved with very high efficiency and charge state determinations and information on fragmentation mechanisms can be obtained.

VI. MASS MEASUREMENT ACCURACY[34,35,52]

A. Introduction

The observed m/z ratio for a given ion often differs from the expected m/z ratio when resonance ejection is used to extend the m/z range. The difference between these m/z ratios is due to a mass shift, which can originate from a variety of sources, but the most important parameters are the frequency and the amplitude of the resonant ejection voltage. For example, Figure 22 displays mass shifts as a function of mass for CsI cluster ions ejected resonantly at $q_z = 0.454$, 0.303, and 0.227 (2X, 3X, and 4X mass range extension). By altering the ejection frequency (q_{eject}), the measured masses differ from those expected and the mass shift becomes more positive as this frequency decreases. Also, the mass shift, at a fixed ejection frequency, decreases as the mass of the ion increases.

To explain the observed mass shifts qualitatively, the following expressions can be used.[53] When an ion is resonated at a fixed RF amplitude, the maximum excursion of the ion in the axial direction can be represented by

$$z_{max} = \left(\frac{b}{2W} \right) \xi \qquad (9)$$

with

$$b = \frac{4V_{res}}{m\Omega_0^2 z_0^2} \qquad (10)$$

where W is the Wronskian,[53,54] ξ is equivalent to $\Omega_0 t/2$ and V_{res} is the amplitude of the resonance voltage. From these equations, it is apparent that the AC voltage required to eject ions at the same rate varies with q_{eject} as

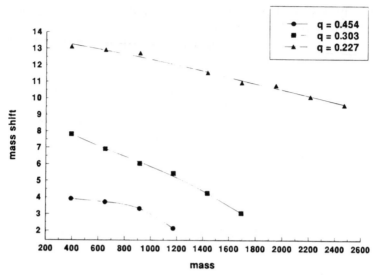

FIGURE 22

Mass shift vs. mass for CsI ions at three values of q_{eject}. (Adapted from Cooks, R.G. and Cox, K.A., Biological Mass Spectrometry: *Present and Future*, Matsuo, T. et al., Eds., John Wiley and Sons, 1994, 179.)

a function of the Wronskian determinant. Equation 10 shows that the amplitude required to eject ions at the same rate (a constant mass shift) is proportional to mass. These two factors match the general trends shown in Figure 22. However, the situation becomes more complicated if we consider the following relationships, which describe the resonance process when the RF voltage is scanned:

$$q = \frac{kV_{RF}}{m} \tag{11}$$

and

$$\frac{dq}{dt} = \frac{k}{m}\frac{dV_{RF}}{dt} \tag{12}$$

Because dV_{RF}/dt is a constant, then dq/dt (or, after further conversions, $d\omega_{z,ion}/dt$) is inversely proportional to mass. This means that the time spent at or near resonance will increase as the mass of the ion decreases. This effect tends to compensate for the requirement of more voltage to eject higher mass ions. However, the magnitudes of these two effects are such that they do not cancel each other, which results in the net (nonlinear) mass shift shown as a function of mass and resonant AC frequency in Figure 22, when the amplitude of the AC voltage is fixed.

Another source for mass shifts arises from variations in the time it takes ions to reach the detector upon ejection from the trap. This time will depend upon the initial conditions of the ions (positions and velocities), perturbations to their trajectories due to space charging and higher-order electric fields, rate of their acceleration to the detector, and collisions with neutrals and other ions.[55,56]

Because of the magnitude of the mass shifts and the fact that they have multiple origins, it is difficult to calibrate the mass scale accurately. This calibration problem is true of low resolution spectra, and the problem becomes magnified in high resolution experiments. In the normal operation of the ion trap, i.e., without enhanced resolution capabilities achieved through slow scans, calibration can be achieved by using CsI cluster ions as external standards. A calibration curve is constructed of the masses observed vs. the known masses of the $(Cs_{n+1}I_n)^+$ cluster ions. The sample is then analyzed using the same resonance ejection voltage and frequency, and mass measurement accuracies of $\leq \pm 0.1$ % result.[27] This accuracy, combined with low mass resolution, is insufficient for differentiation of processes such as water vs. ammonia loss in peptides greater than seven residues. Furthermore, the ionic environment should be identical for both calibrant and sample, which is not the case when external calibration is performed.

B. Internal Calibration of Low Resolution Mass Spectra

In an effort to improve mass measurement accuracy, a series of studies has been conducted using a split probe tip that allows both analyte and calibrant ions to be desorbed by the primary Cs⁺ ion beam and to be injected and trapped simultaneously so as to provide internal mass calibration. To avoid mutual interference during desorption, this tip allows calibrant ions, generated from solids (salt clusters) to be physically separated from the liquid (glycerol) matrix used for peptide analysis. A schematic diagram of the ion trap operated with a split probe tip is shown in Figure 23.

Figure 24 displays a typical mass spectrum (sum of ten scans) obtained by co-injecting CsF and gramicidin S into the ion trap operated at a scan rate of 13,050 Da s⁻¹, which occurs when the m/z range is extended by 2.35 times. Intense CsF cluster ions ranging from m/z 740.5 to 1500.0 and spaced by 151.9 Da are present, along with protonated gramicidin S (shown in greater detail in the inset) and some peptide fragment ions. The inset shows that the resolution in the $[M+H]^+$ region for gramicidin S is 1500 FWHM, which allows each isotopic species to be assigned an m/z value.

Calibration is done by displaying a 10 to 20 Da window and assigning an apparent mass value ±0.05 Da per charge to each ion, using the mass scale provided by the ITMS software (Figure 24, apparent mass scale).

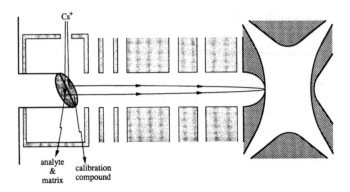

FIGURE 23
Experimental design for co-injection of calibrant and analyte ions.

A calibration plot of the known masses of the CsF cluster ions as a function of their apparent masses is linear and yields a calibrated mass scale (Figure 24, lower scale) from which the gramicidin S ions, including those in the [M+H]+ region, can be assigned m/z values. The calibrated masses are shown for an individual experiment in Table 2 along with the associated mass accuracy values for both CsF and gramicidin S [M+H]+ ions. This experiment was repeated with different ionization times ranging from 3 to 30 ms. An average mass accuracy value of 150 ppm (standard deviation ±90 ppm) is observed in these experiments. From these results, it is clear that calibration using internal standards improves mass measurement accuracy.

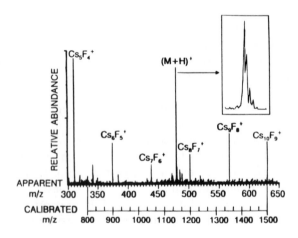

FIGURE 24
Mass spectrum of co-injected CsF and gramicidin S ions. (Adapted from Williams, J.D. and Cooks, R.G., *Rapid Communications in Mass Spectrometry*, Vol. 6, John Wiley and Sons, 1992, 524.)

C. Calibration of Higher Resolution Mass Spectra

The quality of the data, obtained by the method described above, is sufficient for low resolution mass spectra. However, it is desirable to calibrate higher resolution mass spectra to improve further mass assignment. With improved resolution, the sampling across the peak can be increased (see high resolution section) so that the peaks in the isotopic region can be fully resolved and their shapes defined. The method described above cannot, in general, be used to calibrate mass spectra generated by slowing the mass analysis scan. When the previous experiment is performed with the scan rate slowed by a factor of 10, by attenuating the output of the 12-bit mass analysis control DAC, a mass window of m/z 152.9 (i.e., 1/10 of the 1529 m/z range accessed by resonant ejection with a 2.35 mass-range extension factor) is accessible in a single scan. With appropriate positioning of the origin of the scan ramp using an offset voltage, both $Cs_7F_6^+$ (m/z 1044.3) and $Cs_8F_7^+$ (m/z 1196.2) appear in the same scan as the $[M+H]^+$ ions of gramicidin S. However, observing these widely spaced calibrant ions in the same mass window is not generally possible, especially when the scan speed is further slowed.

To overcome problems associated with narrowing of the mass window, calibration is performed by a peak-matching procedure analogous to that used in high resolution mass spectrometers.[57] An offset voltage, which determines where the scan ramp begins, is chosen so that calibrant and analyte ions are in turn brought to the same apparent mass value

TABLE 2

Calibration for Unit Resolution Experiment

Apparent m/z	Theoretical m/z	Calibrated m/z	Δ m/z	Mass accuracy (ppm)
CsF Clusters Used to Calibrate the m/z Scale				
373.7(1)	892.42	892.4(7)	0.05	56
437.7(6)	1044.3	1044.(2)	−0.1	96
501.9(6)	1196.2	1196.(3)	0.1	84
566.0(2)	1348.1	1348.(1)	0.0	—
Protonated Gramicidin S Molecular Ions				
478.9(2)	1141.7	1141.(8)	0.1	88
479.2(7)	1142.7	1142.(6)	0.1	88
479.7(3)	1143.7	1143.(7)	0.0	—
480.2(1)	1144.7	1144.(8)	0.1	88

Calibration curve equation: calibrated m/z = 2.369(3)* (apparent m/z) + 7.(0).

Adapted from Williams, J.D. and Cooks, R.G., *Rapid Communications in Mass Spectrometry*, Vol. 6, John Wiley and Sons, 1992, 524.

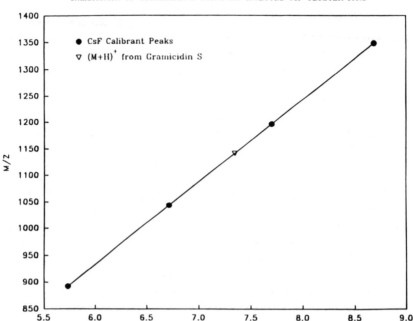

FIGURE 25

Calibration plot produced by peak matching against CsF calibrant ions at enhanced resolution. (Adapted from Williams, J.D. and Cooks, R.G., *Rapid Communications in Mass Spectrometry*, Vol. 6, John Wiley and Sons, 1992, 524.)

(q_{eject} value), which corresponds to a fixed attenuated voltage supplied by the mass analysis control DAC. A calibration line can be constructed by plotting the total applied voltage (offset voltage summed with the attenuated voltage) against theoretical m/z values as shown in Figure 25. As in the previous calibration procedure, a linear response from m/z 892.5 to 1348.1 is obtained.

Table 3 presents calibration results for experiments in which the $[M+H]^+$ region of gramicidin S was recorded at a resolution of 5000 FWHM. Mass accuracies as good as 20 ppm for the third ^{13}C isotope of the protonated gramicidin S peak are achieved, with an average of 50 ppm for all isotopes. When this experiment is repeated using different ionization times, mass accuracies average 70 ppm (standard deviation = 20 ppm) for all isotopic peaks of the protonated molecule. These experiments show the mass measurement accuracy has been improved by a factor of 2 when compared to the low resolution experiment using internal standards. Improvement in accuracy is attributed to both the increased mass reso-

TABLE 3

Calibration for Higher Resolution Experiment

Summed voltage (V)	Theoretical m/z	Calibrated m/z	Δ m/z	Mass accuracy (ppm)
CsF Clusters Used to Calibrate the m/z scale				
5.7335	892.425	892.44(5)	0.020	22
6.1393	1044.33	1044.3(0)	−0.03	29
7.1234	1196.23	1196.2(3)	0.00	—
8.1070	1348.14	1348.1(5)	0.01	7
Protonated Gramicidin S Molecular Ions				
6.7703	1141.72	1141.7(9)	0.07	61
6.7763	1142.72	1142.7(9)	0.07	61
6.7833	1143.72	1143.8(0)	0.08	70
6.7893	1144.73	1144.7(2)	0.01	9

Calibration curve equation: calibrated m/z = 154.37* (summed voltage) + 7.313.

Adapted from Williams, J.D. and Cooks, R.G., *Rapid Communications in Mass Spectrometry*, vol. 6, John Wiley and Sons, 1992, 524.

lution in the slowed scan experiment and to compensation of instrumental errors especially nonlinearities in the RF ramp through the calibration procedure.

D. Calibration of Higher Resolution Tandem Mass Spectra

High mass accuracy is desirable in MS/MS and internal mass calibration of higher-resolution MS/MS spectra has also been performed. A timing diagram for an experiment, which used gramicidin S as the analyte and CsI cluster ions as calibrant, is shown in Figure 26. The supplementary AC in the reverse-then-forward RF portion of the scan (period B) is turned off at several points to allow calibrant ions (m/z 652, 912, and 1172) and the protonated molecular ions of gramicidin S to be trapped. Note that at this stage in the experiment all ions of $m/z < 650$ have been eliminated from the trap. Dissociation of higher mass clusters is performed to regenerate the lower mass cluster ions (stage C); approximately 75% of the $Cs_4I_3^+$ are dissociated to produce $Cs_3I_2^+$ and Cs_2I^+ to provide calibrant ions down to m/z 392. A mass spectrum taken after the isolation period and this dissociation period is shown in Figure 27. Gramicidin S protonated molecular ions are excited resonantly to produce fragment ions during the second excitation step (step D). Figure 28 displays the results recorded under low resolution conditions from this co-injection MS/MS experiment.

Scan Sequence for Co-Injection

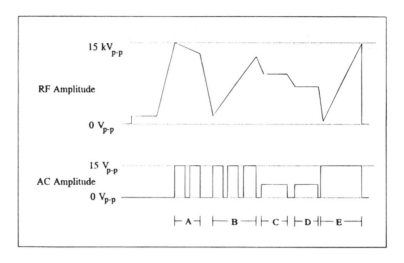

FIGURE 26
Timing diagram for co-injection MS/MS. (Adapted from Cooks, R.G. et al., *Methods in Protein Sequence Analysis*, Inabori, K. and Sakiyama, F., Eds., Plenum, 1993, 135.)

The high resolution MS/MS experiment is performed by slowing the mass-selective instability scan rate (Figure 26, step E) from 11,000 to 1100 Da s^{-1} to provide better than unit resolution throughout the observed m/z range. Mass assignments are then made using the peak matching tech-

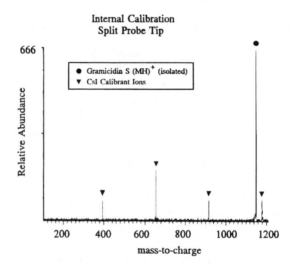

FIGURE 27
Mass spectrum of co-injected gramicidin S and CsI clusters after isolation and dissociation of Cs_4I_3.

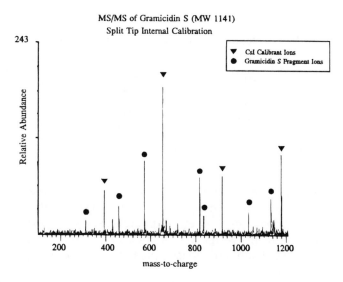

FIGURE 28

MS/MS spectrum of gramicidin S with co-injected CsI cluster ions. (Adapted from Cooks, R.G. et al., *Methods in Protein Sequence Analysis*, Imabori, K. and Sakiyama, F., Eds., Plenum, 1993, 135.)

nique. Instead of using a single calibration line, a moving two-point calibration is used. Three linear functions are created which relate known m/z values for adjacent CsI cluster ions to the total applied voltage that controls the amplitude of the RF at which ion ejection occurs. A moving calibration line improves mass measurement accuracy because mass shifts over a small m/z region can be accounted for more precisely by the slope and intercept of each individual calibration function. Also, this methodology allows parameters such as resonance ejection and frequency to be altered to maximize resolution and signal intensity for each calibration region. The results of this experiment are shown in Table 4. All fragment ions have assigned masses within ± 0.05 Da (≤ 82 ppm) of their theoretical masses.

E.   Future

Large strides have been made to improve mass measurement accuracies in the ion trap, but much more work needs to be done. One feature under study involves examination of the spacings between carbon isotopes of higher mass peptide ions. When the scan rate is slowed, unequal spacings occur (cf. Figure 17. Typically, the ejection of the all-^{12}C isotope occurs later than expected, thus reducing the spacing between it and the $^{13}C_1$ isotope. This situation is complicated further by the presence

TABLE 4

Calibration for Monoisotopic (^{12}C) Gramicidin S Fragment Ions

Theoretical m/z	Calibrated m/z[a]	$\Delta\ m/z$	Mass accuracy (ppm)
571.3608	571.31	0.0468	82
813.5026	813.55	−0.0494	60
831.5132	831.51	0.0015	2
1027.634	1027.6	0.0260	25
1141.715	1141.7	−0.0040	4

[a] Values were calculated using a moving two-point calibration.

of an ion that has an m/z value between the m/z values of the all ^{12}C and ^{13}C$_1$ isotopes as was shown in Figure 17. Table 5 displays the theoretical m/z values of the ions, the theoretical differences, and the observed differences for experiments with and without $Cs_9F_8^+$ present. Clearly, the introduction of the CsF ion into the isotopic pattern has a significant effect on the precise ion spacings.

These mass shifts under discussion are small (< 0.2 Da) compared to those observed previously in low resolution resonant ejection experiments. There are two possible origins for these more subtle effects. Space charging, which is analogous to adding a slight positive potential to the trap, moves the resonance point to higher q_z values, thus requiring higher RF values for ion ejection. Also, prior to the application of the excitation voltage, the ion clouds of lower m/z ions lie closest to the center of the trap,

TABLE 5

Influence of Space Charge on Ions Separated of ≤1 Da/Charge

Isotope (M + H)+	Theoretical $\Delta m/z$	Experimental $\Delta m/z$
Injection of Substance P		
$^{12}C_{63}$		
$^{12}C_{62}{}^{13}C$	1.00	0.9(6)
$^{12}C_{61}{}^{13}C_2$	1.00	1.0(3)
$^{12}C_{60}{}^{13}C_3$	1.00	1.0(3)
Co-Injection of CsF and Substance P		
$^{12}C_{63}$		
$Cs(CsF)_8^+$	0.40	0.3(0)
$^{12}C_{62}{}^{13}C$	0.60	0.4(4)
$^{12}C_{61}{}^{13}C_2$	1.00	1.1(1)
$^{12}C_{60}C_3$	1.00	0.9(8)

due to the shallower pseudo-potential well for the higher mass ions. Upon excitation of their motion by the supplementary AC voltage, ions of lower m/z must travel through the localized clouds of higher m/z ions to be ejected. When the scan speed is slowed, the impediment due to ion/ion interactions becomes increasingly significant. In combination, these effects promote anomalous ejection of abundant ions that are of similar mass.[56]

VII. CONCLUSION

In this chapter various efforts to extend the m/z range, increase mass resolution, and improve mass measurement accuracy were discussed. Development of these methods has greatly enhanced the performance and capabilities of the quadrupole ion trap. Along with these improvements, knowledge of how ions are ejected from the trap, and of the motion of trapped ions has been gained from these studies. While several immediate improvements such as regulating the stability of the RF control circuit and enhancing the data collection process will surely result in enhanced high resolution and mass accuracy experiments, a great deal of effort is still needed to understand better the fundamentals of the motion of trapped ions, the ejection process, the effects of space charge, and both ion/ion and ion/neutral interactions. Simulation studies, as discussed in detail in Volume I, Chapter 6, have already provided valuable insight into ion behavior and have served as a powerful complement to experimental studies. This is particularly so in respect to recent demonstrations[58] of improved resolution and mass measurement accuracy through phase locking the RF and AC signals. Advances in the optimization of the quadrupole ion trap will depend, perhaps, upon correlating theoretical ion behavior mapped by computer simulations with experimental observation.

ACKNOWLEDGMENTS

The contributions to this work by Raymond E. Kaiser, Jr., currently at Eli Lilly, are deeply appreciated. We thank also John Syka, George Stafford, Mark Bier, and John Louris from Finnigan MAT and Phil Hemberger from Los Alamos for their contributions to the progress reported here. The efforts of Kenny Morand, Randy Julian, Jon Amy, and Steve Lammert from Purdue University are appreciated as well. This work is supported by the National Science Foundation, CHE 92-23791. J. D. W. thanks AMOCO Corp. for a Fellowship, and K. A. C. acknowledges a Fellowship from the Analytical Division, the American Chemical Society.

REFERENCES

1. Macfarlane, R.D.; Bunk, D.; Mudgett, P.; Wolf, B. *Mass Spectrometry of Peptides*. D.M. Desiderio, Ed.; CRC Press: Boca Raton, FL, 1991; Chap. 1.

2. Aberth, W.; Straub K.M.; Burlingame, A.L. *Anal. Chem.* 1982, 56, 2029.

3. Barber, M.; Bordoli, R.S.; Sedgwick, R.D.; Tyler, A.N. *J. Chem. Soc. Chem. Commun.* 1981, 325.

4. Karas, M.; Bachmann, D.; Bahr U.; Hillenkamp, F. *Int. J. Mass Spectrom. Ion Processes.* 1987, 78, 53.

5. Fenn, J.B.; Mann, M.; Meng, C.K.; Wong, S.F.; Whitehouse, C.M. *Mass Spectrom. Rev.* 1990, 9, 37.

6. McCloskey, J.A., Ed., *Methods in Enzymology*. Vol. 193. Academic Press: San Diego, CA, 1990.

7. Lambert, J.B.; Shurvell, H.F.; Lightner, D.; Cooks, R.G. *Introduction to Organic Spectroscopy*. Macmillan: New York, 1987; p. 318.

8. Brunée, C. *Int. J. Mass Spectrom. Ion Processes.* 1991, 106, 79.

9. Grix, R.; Kutscher, R.; Li, G.; Gunner, U.; Wollnik, H. *Rapid Commun. Mass Spectrom.* 1988, 2, 83.

10. Lebrilla, C.B.; Wang, D.T.-S.; Hunter, R.L.; McIver, R.T., Jr. *Anal. Chem.* 1990, 62, 878.

11. Beu, S.; McLafferty, F.W. personal communication, 1992.

12. Stafford, G.C.; Kelley, P.E.; Syka, J.E.P.; Reynolds, W.E.; Todd, J.F.J. *Proc. 13th Meet. Br. Mass Spectrom. Soc.* Warwick, U.K., 1983; p. 18.

13. Stafford, G.C., Jr.; Kelley, P.E.; Syka, J.E.P.; Reynolds, W.E.; Todd, J.F.J. *Int. J. Mass Spectrom. Ion Processes.* 1984, 60, 85.

14. Louris, J.N.; Cooks, R.G.; Syka, J.E.P.; Kelley, P.E.; Stafford, G.C.; Todd, J.F.J. *Anal. Chem.* 1987, 59, 1677.

15. Louris, J.N.; Amy, J.W.; Ridley, T.Y.; Cooks, R.G. *Int. J. Mass Spectrom. Ion Processes.* 1989, 88, 97.

16. Kaiser, R.D., Jr.; Williams, J.D.; Lammert, S.A.; Cooks, R.G.; Zakett, D. *J. Chromatogr.* 1991, 562, 3.

17. Bier, M.E.; Hartford, R.E.; Herron, J.R.; Stafford, G.C. *Proc. 39th ASMS Conf. Mass Spectrom. Allied Topics*. Nashville, TN, 1991; p. 538.

18. Van Berkel, G.J.; Glish, G.L.; McLuckey, S.A. *Anal. Chem.* 1990, 62, 1284.

19. Jardine, I.; Hail, M.E.; Lewis, S.; Zhou, J.; Schwartz, J.C. *Proc. 38th ASMS Conf. Mass Spectrom. Allied Topics*. Tucson, AZ, 1990; p. 16.

20. Voyksner, R.D.; Lin, H.Y. *Proc. 39th ASMS Conf. Mass Spectrom. Allied Topics*. Nashville, TN, 1991; p. 526.

21. McLuckey, S.A.; Van Berkel, G.J.; Glish, G.L.; Huang, E.C.; Henion, J.D. *Anal. Chem.* 1991, 63, 375.

22. Kaiser, R.E., Jr.; Louris, J.N.; Amy, J.W.; Cooks, R.G. *Rapid Commun. Mass Spectrom.* 1989, 3, 225.

23. Schwartz, J.C.; Bier, M.E. *Rapid Commun. Mass Spectrom.* 1993, 7, 27.

24. Jonsher, K.; Currie, G.; McComack, A.L.; Yates, J.R., III *Rapid Commun. Mass Spectrom.* 1993, 7, 20.

25. Chambers, D.M.; Goeringer, D.E.; McLuckey, S.A.; Glish, G.L. *Anal. Chem.* 1993, 65, 14.

26. Kaiser, R.E., Jr.; Cooks, R.G.; Moss, J.; Hemberger, P.H. *Rapid Commun. Mass Spectrom.* 1989, 3, 50.

27. Kaiser, R.E., Jr.; Cooks, R.G.; Stafford, G.C., Jr.; Syka, J.E.P.; Hemberger, P.E. *Int. J. Mass Spectrom. Ion Processes.* 1991, 106, 79.

28. Schwartz, J.C.; Syka, J.E.P.; Jardine, I. *J. Am. Soc. Mass Spectrom.* 1991, 2, 198.

29. Williams, J.D.; Cox, K.A.; Cooks, R.G.; Kaiser, R.E., Jr.; Schwartz, J.C. *Rapid Commun. Mass Spectrom.* 1991, 5, 327.

30. Schwartz, J.C.; Jardine, I. *Rapid Commun. Mass Spectrom.* 1992, 6, 313.
31. Cooks, R.G.; Cox, K.A.; Williams, J.D.; Morand, K.L.; Julian, R.K., Jr.; Kaiser, R.E., Jr. *Proc. 39th ASMS Conf. Mass Spectrom. Allied Topics.* Nashville, TN, 1991; p. 469.
32. Cox, K.A.; Williams, J.D.; Cooks, R.G.; Kaiser, R.E., Jr. *Biol. Mass Spectrom.* 1992, 21, 226.
33. Londry, F.A.; Wells, G.J.; March, R.E. *Rapid Commun. Mass Spectrom.* 1993, 7, 43.
34. Williams, J.D.; Cooks, R.G. *Rapid Commun. Mass Spectrom.* 1992, 6, 524.
35. Cooks, R.G.; Cox, K.A.; Williams, J.D. *Methods in Protein Sequencing.* F. Sakiyama, Ed. Plenum Press: New York, 1993, 135.
36. Knight, R.D. *Int. J. Mass Spectrom. Ion Phys.* 1983, 51, 127.
37. Stafford, G.C., Jr.; Syka, J.E.P. unpublished results.
38. Todd, J.F.J.; Penman, A.D.; Smith, R.D. *Int. J. Mass Spectrom. Ion Processes.* 1991, 106, 117.
39. Fischer, E. *Z. Phys.* 1959, 156, 1.
40. Syka, J.E.P.; Schwartz, J.C.; Louris, J.N.; Amy, J.W.; Fies, W.J., Jr.; Fenske, S.A. *Proc. 39th ASMS Conf. Mass Spectrom. Allied Topics.* Nashville, TN, 1991; p. 544.
41. Weber-Grabau, M.; Kelley, P.E.; Bradshaw, S.C.; Hoekman, D.J. *Proc. 35th ASMS Conf. Mass Spectrom. Allied Topics.* Denver, CO, 1987; p. 263.
42. Paul, W.; Reinhard, H.P.; von Zahn, V. *Z. Phys.* 1958, 152, 143.
43. Young, S.E. *Proc. 40th ASMS Conf. Mass Spectrom. Allied Topics.* Washington, D.C., 1992; p. 1759.
44. Goeringer, D.E.; McLuckey, S.A.; Glish, G.L. *Proc. 39th ASMS Conf. Mass Spectrom. Allied Topics.* Nashville, TN, 1991; p. 532.
45. March, R.E. *Int. J. Mass Spectrom. Ion Processes.* 1992, 118/119, 71.
46. Williams, J.D.; Cox, K.A.; Morand, K.L.; Cooks, R.G.; Julian, R.K., Jr. *Proc. 39th ASMS Conf. Mass Spectrom. Allied Topics.* Nashville, TN, 1991, p. 1481.
47. Nedderman, A.N.R.; Williams, D.H. *Biol. Mass Spectrom.* 1991, 20, 289.
48. Morand, K.L.; Reiser, H.-P.; Williams, J.D.; Julian, R.K.; Cooks, R.G. *Proc. 40th ASMS Conf. Mass Spectrom. Allied Topics.* Washington, D.C., 1992; p. 1779.
49. Reiser, H.-P.; Julian, R.K., Jr.; Cooks, R.G. *Int. J. Mass Spectrom. Ion Processes.* 1992, 121, 49.
50. Julian, R.K.; Reiser, H.-P.; Cooks, R.G. *Int. J. Mass Spectrom. Ion Processes.* 1993, 123, 85.
51. Goeringer, D.E.; Whitten, W.B.; Ramsey, J.M.; McLuckey, S.A.; Glish, G.L. *Anal. Chem.* 1992, 64, 1434.
52. Cooks, R.G.; Cox, K.A. *Recent Developments in Biological Mass Spectrometry:* Present and Future, T. Matsuo, M.L. Gross, and R.M. Caprioli, Eds. John Wiley and Sons, 1994, 179.
53. Dawson, P.H. *Quadrupole Mass Spectrometry and Its Applications.* Elsevier: Amsterdam, 1976; p. 49.
54. March, R.E.; Hughes, R.J. *Quadrupole Storage Mass Spectrometry.* Chemical Analysis Series, Vol. 102. Wiley Interscience: New York, 1989; p. 74.
55. Reiser, H.-P. Ph.D. thesis, Justus-Liebig-Universitat, Giessen, Germany, Chapter 4.
56. Cox, K.A., Cleven, C.D., and Cooks, R.G., *Int. J. Mass Spectrom. Ion Proc.,* in press.
57. Beynon, J.H. *Mass Spectrometry and Its Application to Organic Chemistry.* Elsevier: Amsterdam, 1960; p. 38.
58. Cleven, C.D. and Cooks, R.G., unpublished results.

PART 2

Ion Trap Confinement of Externally Generated Ions

Chapter 2

INJECTION OF MASS-SELECTED IONS INTO THE RADIOFREQUENCY ION TRAP

Peter Kofel

CONTENTS

I.   Introduction 52

II.   Principal Aspects 53
     A.   Trapping 53
         1.   Kinetic Energy Reduction by Electrostatic
             Deceleration 55
         2.   Kinetic Energy Reduction by a Dissipative
             Process 55
         3.   Potential Energy Reduction by Pulsed
             Operation of the Trap 55
     B.   Injection 56
     C.   Mass Selection 58

III.   Deceleration of Fast Ions 58

IV.   Comparison of Ion Injection into Paul Trap and
      Penning Trap 62

V.   Instrumental Realizations 63
     A.   Mass Selection by the Radiofrequency Ion Trap 64
     B.   Mass Selection by a Magnetic Sector 66
     C.   Mass Selection by Quadrupole 68

0-8493-4452-2/95/$0.00+$.50
© 1995 by CRC Press, Inc.

VI.   Quadrupole, Quistor, Quadrupole Instrument 69
       A.   Description of the Instrument 69
       B.   Description of the Ion Optics 70
       C.   Injection Efficiency 73
       D.   Trapping Process 73
       E.   Relaxation Process 76
       F.   Trapping Efficiency 79
       G.   Pressure in the Ion Trap 80
       H.   Synchronization 81
       I.   Acquisition 82
       J.   Studies of Ion/Molecule Reactions 83

VII.   Conclusions 85

Acknowledgment 85

References 86

I. INTRODUCTION

The radiofrequency (RF) quadrupole ion trap[1] is a fascinating device due to its versatility combined with its mechanical simplicity. With only a filament and a secondary electron multiplier attached to it, the trap can be used as a complete mass spectrometer.[2] The same physical volume, confined by only the three electrodes of the trap which are a ring electrode and two end-cap electrodes, may be used as an ion source, for ion storage, and for mass analysis. The ion storage period may be used for virtually any desired interaction with the stored ions such as mass selection, ion activation by collisions or lasers, ion fragmentation by collision-induced dissociation (CID), surface-induced dissociation (SID), photodissociation (PD), and ion/molecule reactions.[3-5] The ability to utilize long interaction times and to achieve high product ion collection efficiencies are desirable characteristics of the ion trap.[6] To make the ion storage period accessible to all kinds of applications, it is often desirable to be able to store ions injected into the trap. Typically the applications that require injection of ions into the trap are those in which ions are formed under conditions of high gas pressure. The removal of ions from the high pressure region and storage within the ion trap is both a requirement for trap operation (in that pressures in excess of 10^{-1} Pa are incompatible with high RF voltage, mass analysis, and detection) and a method whereby unwanted neutrals that may interfere with the experiments during the storage period are removed.

For many of the applications involving injection of ions into a trap it is also desirable to exclude a certain range of m/z ratios or even all but a single m/z ratio from being injected or trapped or both. The former is the case in which trapping of all ions would fill the trap to the space-charge limit before a sufficient number of low relative abundance ions of interest can be trapped. A typical application is the exclusion of the low mass matrix or solvent ions present from high mass ionization techniques. A single m/z ratio injected often facilitates or even only makes possible the interpretation of the experiments performed within the trap. Mass-selected injection is therefore used typically when low abundance ions are to be accumulated in the trap for subsequent characterization by dissociation or ion/molecule reaction experiments.

This chapter focuses on injection of mass-selected ions into RF ion traps. Discussion of the principal aspects of mass selection and trapping of injected ions, including the results of numerical studies on ion injection in Section II, are followed by an exposition on the deceleration of fast ions in Section III and a comparison of injection of ions into Paul traps and Penning traps (magnetic ion trap) in Section IV. Instrumental realizations of mass-selected injection of ions into quadrupole ion traps are reviewed and discussed in Section V and, as an example, the quadrupole, quistor, quadrupole instrument is discussed in more detail in Section VI.

II. PRINCIPAL ASPECTS

Trapping of ions injected into the Paul trap has been the subject of a number of theoretical studies, reviewed by March and Hughes.[3] Since then, several new experimental reports on ion injection have appeared in the literature. The main findings of these studies are included in this section.

A. Trapping

One of the basic parameters for the trapping of injected ions is the choice of working point in the (a_z, q_z)-stability diagram shown in Figure 1. It is necessary that the working point lies within the region for stable trapping.[3] When the trap is operated in the RF-only mode, this condition for stable trapping requires that the RF amplitude be chosen such that the value of q_z lies within the stability boundary at a value less than $q_z = 0.908$. In addition, a strong dependence of injection efficiency upon the position of the working point within the stability region has been found. These effects are discussed in detail in Section II.C. For the following discussion on trapping it is assumed that the ion trap is set for stable trapping of the ion species under consideration.

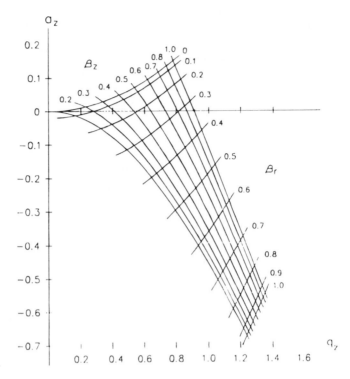

FIGURE 1
Stability region for the three-dimensional radiofrequency ion trap. (From Reference 3; reproduced by permission of John Wiley & Sons, Ltd.)

Under continuous trapping fields with constant amplitudes and in the absence of ion motion relaxation, trapping of a particle injected into the trap is impossible by definition. This is because the definition of trapping implies that the particle must return to a position arbitrarily close to the starting point. The starting point of an injected particle is outside the trap boundaries, so the particle will eventually leave the trap boundaries again. This behavior has been recognized from the beginning of Paul trap experiments.[7]

However, it may take a long time for the particle to come back to its starting point. Numerical calculations indicate that substantial path lengths can be found for ions injected in narrow initial RF phase windows.[8] Depending on the initial conditions an ion may be inside the trap for many cycles of the RF field.[9]

In its trajectory through the RF field, the ion will retain an excess amount of (injection) energy. The excess energy will allow the ion to reach the trap boundaries again, at the latest, when one trajectory cycle is closed.

Thus, in order to store an ion for a longer period, the excess energy must be removed in some way. Three principal possibilities exist by which this objective may be achieved.

1. Kinetic Energy Reduction by Electrostatic Deceleration

Electrostatic deceleration can be employed outside the trap: as the ions approach the trap, they are slowed down to a few electronvolts.[10,11] Although electrostatic deceleration helps to reduce the kinetic energy of the ions, they cannot be trapped by this method alone. Deceleration is discussed further in Section III.

2. Kinetic Energy Reduction by a Dissipative Process

Dissipative processes can be used for both slowing down approaching ions (even kiloelectronvolt ions[12]) and final trapping of the ions.[13,14] A number of relaxation techniques have been proposed and examined, such as radiative coupling,[15] ion/molecule collisions,[16,17] ion/surface collisions,[11] ion/bulk collisions,[9] and motional sideband cooling.[18] Ion energy manipulation in RF ion traps has been discussed in detail recently.[19] Here, only the process of trapping injected ions by ion/molecule collisions is discussed because it seems to be of the broadest applicability for mass spectrometry. Trapping under collisional conditions can be characterized as follows. An ion should enter the trap with low translational energy; while the ion passes through the trap, excess translational energy must be removed by collisions with neutrals, which is best achieved in collisions with neutrals of the same mass. However, if the ion has lost enough energy that it can no longer escape the trap, additional collisions should further relax its motion in the RF field to a small region close to the center of the trap; such collisional focusing requires a gas of mass lower than that of the ion.[17] In practice, a compromise must be made. When only one gas is used, helium has been found to be usually superior to the other rare gas targets, although it is expected that for very heavy ions a heavier target may be more effective.[14] An alternative compromise would be the use of buffer gas mixtures. Small amounts of Ar mixed with He have shown superior trapping efficiency for $Cs(CsI)_{31}^+$ cluster ions (m/z 8187).[20]

3. Potential Energy Reduction by Pulsed Operation of the Trap

The pulsed technique can be used for ions already in the trap. By an instantaneous change in the operating conditions, the well depth of the trap is increased and ions of low kinetic energy close to the center of the trap can no longer escape. Depending on the working point in the stability diagram and the initial RF phase, ions can be trapped from a considerable phase space volume, defined by initial ion velocities and exact ion locations at the onset of the trapping voltage.[21-23] For potential isolation purposes, a pulsed technique also could be employed with an "ion elevator" in front of the trap, similar to that described by Smalley and co-workers[24] for their external source Fourier transform-ion cyclotron resonance (FT-ICR) instrument.

It has been assumed that the pulsed trapping could be achieved only by pulsing the RF trapping potential. However, it must be pointed out here that any method in which the potential well depth is increased can be used. As an example, when a trap is operated with the RF potential on the ring electrode only, trapping of injected ions can be achieved with appropriate DC pulses applied to one or both end-cap electrodes. When a negative pulse is applied to the entrance end-cap electrode during ion injection, positive ions that are in the trap at the end of the pulse can be trapped; this procedure is the inverse of the pulse-out detection method. Similarly, positive ions can be trapped when a negative pulse is applied to both end-cap electrodes, due to a sudden change of the working point in the stability diagram, as is used for ion isolation purposes. Compared to RF pulsing, fast DC pulsing is straightforward; it has been applied successfully to the capture of fragment ions produced by SID when parent ions are made to collide with one of the ion trap electrodes.[25] It may be easily applicable to laser desorption (LD) experiments, in which ions are produced in a pulsed fashion at the trap boundary[26] or outside the trap.[14]

B. Injection

Injection (transmission) of ions into an RF ion trap is constrained by the demands that a large portion of the ion beam be transmitted to the central region of the trap and that the trapping field should not be disturbed too much. The second constraint means in essence that the entrance hole must be made small. Injection of the ions through the gap between the electrodes would require no additional holes (Figure 2). For pulsed operation of the trap it was found theoretically that injection along the asymptote between ring and end-cap electrodes would be most favorable, and axial injection through one of the end-caps least efficient.[23] Therefore, pulsed trapping would allow one to inject the ion beam efficiently along the asymptote.

In contrast to the latter results, it was found that for a continuous trapping field, injection from an end-cap is most favorable and that low energy ions with a high incidence angle have the longest residence time in the trap (Figure 3).[9] Ions injected through the asymptotic gap would

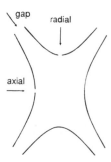

FIGURE 2
Quadrupole ion trap in cross-section showing three different injection situations.

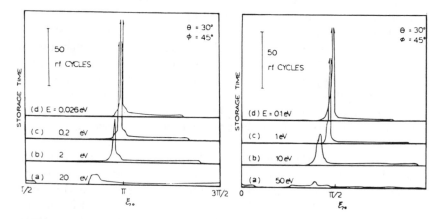

FIGURE 3
Storage time (measured in RF cycles) as a function of initial RF phase of an ion beam injected through the midpoint of the ring electrode (left) and through the apex of one of the end-cap electrodes (right) for favorable incidence directions ($\theta = 30\,°$, $\phi = 45\,°$) and different energies. (From Reference 9; reproduced by permission of the American Institute of Physics.)

not survive a single RF cycle. For injection through an electrode, a small hole is a major limitation for efficient injection because low energy ions are required for efficient trapping. However, depending on the characteristics of the injected ion beam and the amount of deviation from pure quadrupole potential tolerated, alternative ion entrance designs may be proposed (see Chapter 8). For the injection of an ill-defined ion beam, such as that emerging from a quadrupole, it might be useful to have multiple holes in the entrance electrode, similar to the design of the exit end-cap electrode in a commercial ion trap,[4] or to use a larger hole covered with a piece of wire mesh approximating the hyperbolic shape. If exact quadrupolar geometry is not required, but only the realization of an ion storage capability, the entrance hole may be enlarged, or cylindrical ion trap geometry[27] may be chosen, or the geometry may be extended to quite open designs of concentric cylinders.[28,29]

C. Mass Selection

Mass selection of an injected ion beam can be achieved either outside or inside the trap. Whereas for the mass selection outside the trap any beam mass analyzer can be employed (see Section V), the methods for mass selection inside the trap depend on the trapping method used. In the pulsed trapping mode, mass selection could be achieved by a time-of-flight (TOF) selection. The mass of interest is chosen by varying the time delay between ionization and the trapping pulse.

In the collisional trapping mode, the m/z ratio of the ion species to be trapped can be selected, in principle, by the choice of the stability parameters a_z and q_z of the trap during the injection time. However, the closer the working point is to a stability limit for radial or axial motion, the more difficult it is to confine the ion motion in the corresponding direction.[9] For the most efficient trapping, the best working point appears to be close to the point where the potential well is deepest (around $a_z = 0$ and $q_z = 0.5$).

In addition to the latter effects, which apply equally well to ions generated within the trap, two further observations have been made regarding the trap working point during ion injection. First, when the working point is chosen such that the ion oscillates at a subharmonic of the RF trapping field frequency, its trajectory cycle is closed very quickly and it may strike the physical boundaries of the ion trap before being slowed down.[8] Subharmonic oscillation in the direction of injection is, therefore, expected to have a negative effect on trapping efficiency. Such resonance effects have indeed been observed for $q_z = 0.635$, where the ion oscillates at one fourth of the trapping field frequency.[30] These resonance effects were explored in more detail in a subsequent study.[31] Second, the optimal RF level for efficient trapping has been found to increase both with ion mass[14,32] (Figure 4) and injection energy.[30]

III. DECELERATION OF FAST IONS

The transport of a charged particle beam in static electromagnetic fields is governed by the Lagrange-Helmholtz equation:[33]

$$\Delta\rho \, \Delta v_r = (\Delta\rho)_0 (\Delta v_r)_0 \tag{1}$$

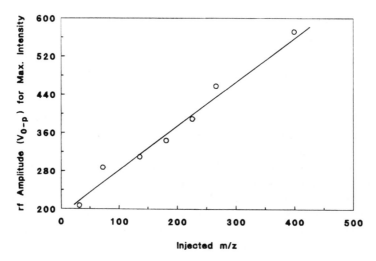

FIGURE 4
RF voltage amplitude at which maximum injection efficiency is observed as a function of ion mass. (From Reference 32; reproduced by permission of Elsevier Science Publishers B.V.)

which states that the product of the widths of radii, $\Delta\rho$, and radial velocities, Δv_r, of the ion beam in waist-to-waist transport is constant. This equation is more commonly expressed as

$$\Delta\rho \sin \Theta E_{\text{kin}}^{1/2} = \text{const} \tag{2}$$

where E_{kin} is the kinetic energy of the ions and Θ is the opening angle of the beam. When the ions are produced with thermal initial velocities within a disc of radius r_0, the opening angle Θ is given by

$$\sin \Theta = \frac{r_0}{\Delta\rho} \sqrt{\frac{k_B T}{E_{\text{kin}}}} \tag{3}$$

where k_B is the Boltzmann constant and T is the ion source temperature. Equation 3 is an idealized case, usually strongly underestimating the opening angle. It is strictly correct only for thermal desorption of ions from a hot surface, but it still may be used for the illustration of the effects encountered in dealing with slow ions. While the functional dependence of Equation 2 is still valid for other ion sources, the constant is larger.

In practice, the maximum allowed opening angle is defined by the geometry of the ion optics. A lens system designed for large opening angles must have either several focusing elements or the lens electrodes must have much wider inner diameters to transport the beam without losses over a given distance. Often the size of a hole is given through which the beam must pass. Such is the case when the beam must pass through gas flow restrictions, resolution-defining slits, or field-defining electrodes. An enhanced kinetic energy is then often used to achieve high transmission. However, when low energy transfer is required, the only parameter that can be optimized is the opening angle. Yet the opening angle cannot be more than the theoretical limit of 90°. Therefore, even with an opening angle close to this limit, a compromise must be found among small hole size, low injection energy, and high transmission.

For low energy ion beams, the energy spread of the ions becomes an important parameter. First, the spread in kinetic energy is a limit for deceleration. When ions of lower kinetic energy are required, the slower ions are reflected in the deceleration stage, and beam transmission is reduced. From this point of view, the combination electrostatic sector/deceleration is potentially advantageous over the combination quadrupole/deceleration. In practice, the energy spread also translates to an increased beam waist size at low energies.

Because deceleration has the effect of widening the ion beam, a deceleration lens system should be open and allow for refocusing the diverging beam. A potential exists that has both properties, deceleration and continuous refocusing of the ion beam, where an explicit solution for the ion trajectories can be given. It is an exponentially decaying potential given by

$$\Phi(\rho, z) = -V_0 e^{-x_1 z/r} J_0\left(x_1 \frac{\rho}{r}\right) \tag{4}$$

where z is the distance along the axis, ρ is the distance from the axis, $J_0(x)$ is the Bessel function of order zero, $x_1 = 2.40483$ is the first zero of $J_0(x)$ ($J_0(x_1) = 0$), r is a scaling length, and $-V_0$ is a voltage ($\Phi(0, 0) = -V_0$). The use of this potential for deceleration has long been recognized and many decelerator lens systems were constructed using a stack of axially symmetric plates connected to an exponentially decreasing resistive voltage divider.[34] Only recently has careful analysis of the exponential potential led to a much simpler design consisting of only two electrodes for an exponential deceleration lens[35] (Figure 5). A similar design with a flat entrance electrode and a cylinder is successfully used for deceleration and injection of ions into magnetic ion traps.[35]

Moreover, deceleration of energy-selected ions has been achieved successfully with a nonexponential design,[36] and high transmission has been reported.

Because RF quadrupolar fields themselves have refocusing properties, alternative deceleration designs may be chosen for the injection of an ion beam into a quadrupole or an RF ion trap. Deceleration may not have to be as gentle and controlled when deceleration is effected at the entrance to these devices, and refocusing is guaranteed by the immediately following RF fields.

Two simple designs for quadrupole and RF ion trap requiring no additional voltages except those for focusing to the entrance hole are illustrated in Figure 6. These designs are mechanically simple and ensure a high transmission of ions and low final ion energy. They are used for commercial quadrupoles[37] and for ion injection into an RF ion trap[38] (see

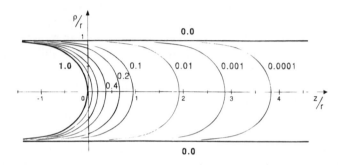

FIGURE 5
Simple deceleration lens producing an exponentially decreasing potential. The two electrodes are drawn in thick lines. The equipotential values given are normalized by $-V_0$. Ion injection is from the left through a small hole in the center of the hemispherical-like electrode. The radius of curvature of the equipotentials at the axis is $R = 0.83166\ r$. (From Reference 35; reproduced by permission of Elsevier Science Publishers B.V.)

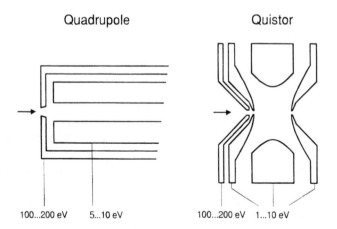

FIGURE 6
Final deceleration stage for the injection of ions into a quadrupole mass filter (left) and an RF quadrupole ion trap (right).

Section VI), respectively. For the deceleration of ions in the kiloelectron-volt regime, a predeceleration stage of one of the designs discussed above must be employed.

IV. COMPARISON OF ION INJECTION INTO PAUL TRAP AND PENNING TRAP

In a Paul trap, ions are trapped in both radial and axial directions by the RF electrical field. In a Penning trap, ions are confined radially by a magnetic field and trapped axially by a DC electrical field. Penning traps with resonant mass detection are used in ICR mass spectrometry, which has been reviewed recently.[39] Whereas for the Paul trap a low mass trapping limit is defined by the boundary of the Mathieu stability diagram,[3] for the Penning trap a high mass trapping limit exists because of the radial electric field component associated with the axial trapping field.[39] With regards to collisions with neutrals, ions are stabilized in the Paul trap by collisions when the neutral mass is smaller than the ion mass, and destabilized when the neutral mass is larger than the ion mass.[17] In a Penning trap, collisions of ions with neutrals result in radial diffusion of the ions, which ultimately limits the trapping time.[40,41] Attempts have been made in the field of ICR spectrometry to overcome both the radial electric field with screened cells[42] and the collision number limit by applying additional RF electric fields during high pressure events.[43] However, an ultimate upper mass limit is operative for both types of traps and is given by

$$m \leq \frac{p_{max}^2}{2k_B T} \tag{5}$$

with $p_{max} = \frac{eV}{\sqrt{2}r_0\Omega}$ for the Paul trap and $p_{max} = eBr_0$ for the Penning trap.

At this limit, the thermal motion of the ions (T is the ion temperature) reaches the trap boundaries.

With respect to ion injection, the obvious difference between the RF trap and the magnetic trap lies in the nature of the respective fields used; electrical fields are shielded effectively by any conducting material, whereas magnetic fields can be shielded only by thick layers of magnetic material. Therefore, the electric RF field of the Paul trap is generated by applying an RF voltage directly to the trap ring electrode, and the field is limited basically to the interior of the trap; the magnetic field used for a Penning trap is generated outside the vacuum system and the magnetic stray field is significant even at the site of an external ion source.

Compared to the injection of ions into an RF ion trap, it is more difficult to bring the ions to the entrance electrode of a magnetic trap because the magnetic mirror effect must be overcome. However, because the magnetic field helps to keep the ions bundled together, it is easier to decelerate the ions and to get them into the magnetic trap. Also, the entrance hole can be made much larger because the electrode geometry can be selected independently from the magnetic field. Trapping of slow ions with pulsed potentials is straightforward with a Penning trap, whereas collisional trapping is of limited use. As the trapping field is not dynamic, there is no major phase dependence when using pulsed trapping. The magnetic field has been taken into account in the design of the external ion source. When a crossed electron beam is used for ionization, the ion source may have to be shielded.[35]

Due to radial bundling, ions can be injected into magnetic traps at translational energies as low as 0.1 eV.[35] The ions then have flight times through the trap on the order of 10^{-4} s. Such flight times allow efficient trapping of fragile ions for subsequent studies in the magnetic trap.[44] Ions are also efficiently thermalized in a magnetic trap after a few collisions with neutrals. In contrast (as shown in Section II.B) it is difficult to inject low energy ions into an RF trap, as ion kinetic energy changes for different RF phases and RF heating of the ions may dissociate fragile ions in ion/neutral collisions.

Various designs are being used for overcoming the magnetic mirror effect and to inject ions into magnetic traps. Quadrupoles for guiding the ions through the field gradient,[45-47] electrostatic focusing of the ions with an Einzel lens to align them to the magnetic field lines and pulsed deceleration with an ion elevator,[24] and a simple cylindrical deceleration/injection lens assembly have been considered.[35,48] Theoretically, an effect similar to the magnetic mirror effect is expected for the injection of charged particles into RF fields. The effect is due to the fringing fields close to the injection hole. For axial or radial injection, the effect can be described by an entrance barrier proportional to the well depth of the RF trap (see Volume I Chapter 2), but reduced by a geometric factor depending on the ratio of entrance hole size to trap size. For constant RF amplitude, low mass ions would have to surmount a higher barrier than high mass ions. So far, this effect has not been systematically studied.

V. INSTRUMENTAL REALIZATIONS

Experimental investigations on mass-selected injection have been reported only for collisional conditions. The results of studies in which the ion mass is selected by the operation conditions of the RF ion trap are included in this section because they allow one to realize the use of mass selection external to the trap.

A. Mass Selection by the Radiofrequency Ion Trap

Cooks and co-workers[14] have explored the feasibility of an external ion source coupled to a commercial ion trap detector in order to improve access to the ionization region. They injected low energy ions (10 to 25 eV) axially into a quadrupole ion trap having a ring electrode radius of 1.0 cm and at an ambient pressure of 2.7×10^{-4} torr of He collision gas. The trapped ions were analyzed by the method of mass-selective instability. For an analyte pressure of 10^{-6} torr of perfluorotributylamine, and injection times somewhat longer than internal ionization times, good signal-to-noise (S/N) ratios were obtained for electron impact ionization. It was found that the trapping efficiency was dependent upon both the mass of the injected ion and the amplitude of the RF voltage during injection

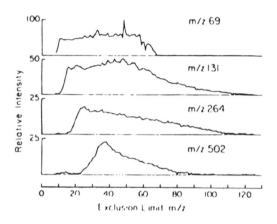

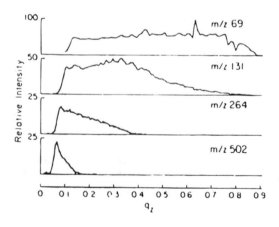

FIGURE 7
Relative abundances of four ions from perfluorotributylamine as a function of RF voltage (top), and the same data plotted as a function of q_z (bottom). Injection energy was 10 eV. (From Reference 14; reproduced by permission of Elsevier Science Publishers B.V.)

(Figure 7). They were able to rationalize the threshold for efficient trapping of an ion to be proportional to $m^{1/2}$ (threshold RF voltage V proportional to $m^{1/2}$). Cooks et al. found also that larger ions are trapped in increasingly narrower q_z ranges, although their threshold q_z is lower (Figure 7, bottom). This behavior was attributed partly to the inefficiency of the damping collisions because of the large disparity in masses. It is interesting to note that this mass-dependent threshold was not observed for ions desorbed by a laser from the trap boundary.[26] The ions injected from the trap boundary behaved similarly to ions produced internally by electron or chemical ionization (EI, CI).

The trapping efficiency of laser-desorbed Au^+ ions with different collision gases was studied. A single laser pulse was found to be capable of filling the trap to the space-charge limit. Nominal pressures (Bayard-Alpert ionization gauge) of at least 1×10^{-5} torr of He were required for trapping. For trapping injected Au^+ ions, neon was found to be as effective as He, but Ar and Xe yielded signals only 10 to 20% as intense as that found for He. For all gases, the highest nominal pressure (2×10^{-4} torr) yielded the most intense signals for trapped ions. Ion/molecule reactions and photodissociation of the injected ions were demonstrated. No account was made in their report for the absolute efficiency of the injection/trapping process. From this study, it was clear that the RF ion trap alone can be used to suppress trapping of certain m/z ratios by selecting the RF voltage as desired. Low masses are not trapped at all if they are below the exclusion limit, and high masses are not trapped efficiently if their threshold is above the exclusion limit. With the same injection optics CsI clusters up to m/z 21´000 have successfully been injected.[49]

Reduced fragmentation upon injection was observed by Yost and co-workers when ions were injected with a lower kinetic energy of 5 eV from a chemical ionization source.[50] Direct neutral gas flow from the ion source to the trap was eliminated by employing a dc quadrupole deflector and ion optics open to gas flow between source and trap. Resonance effects were reported in the injected ion spectra. Higher mass ions (m/z > 200) were observed not to suffer from this phenomenon, because they are optimally injected at low q_z values, where resonance effects are less prevalent. Overall, injection/trapping/detection efficiency was reported to be only 0.03% at 5 eV injection energy, although deceleration of the ions was effected close to the entrance of the ion trap. This low efficiency may be due to the lack of focusing of the ion beam to the entrance electrode.

Many high gas load applications incorporating injection of ions combined with differential pumping between ion source and the ion trap have been reported, and include atmospheric sampling glow discharge,[32] electrospray,[51] particle beam interface, and thermospray interface.[52]

B. Mass Selection by a Magnetic Sector

The first report of mass-selected injection of ions into an RF ion trap was made by March and co-workers.[13] In their experiment, ions were injected from an EB sector instrument into the trap through the gap between ring and end-cap electrodes at an angle of 125° (Figure 8). Ions were injected at their full 4 keV acceleration energy. A collision gas was admitted to the trap through a pulsed valve. Stored ions were extracted through a hole in one end-cap electrode of the ion trap and mass analyzed by an attached quadrupole mass filter. The aim of that experimental setup was to study high energy CID processes, which result in larger scattering angles and are therefore discriminated against in conventional CID collision cell experiments. A Monte Carlo simulation study indicated that about 2% of the parent ions may be trapped intact after having undergone 15 collisions, while >4% of the fragment ions were collected. Two operating modes of the trap were reported: in mode I, the trap was used as an integrating detector for the sector instrument, exploiting the fact that injected ions can be accumulated in the trap; in mode II, the ion trap and quadrupole mass filter served as the second stage for MS/MS analysis.

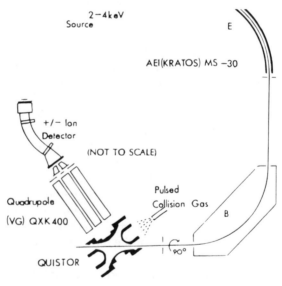

FIGURE 8

Hybrid mass spectrometer in which mass-selected ions of 4 keV energy are directed asymptotically into a pulse-pressurized RF ion trap. Trapped ions are ejected at intervals into a quadrupole mass filter for mass analysis. (From Reference 3; reproduced by permission of John Wiley & Sons, Ltd.)

In an improved version of the instrument,[12] collision gas addition through a pulsed valve was replaced by a constant collision gas flow and a collision cell in front of the trap. The collision cell served as a deceleration stage, while the thermalization step (ion energy <10 eV) had to occur within the trap in order to store an injected ion. An estimated 0.5 to 3% of the total injected ions could be trapped. It was observed that a fraction of the molecular ions survived collisional thermalization intact. It was observed also that the ion beam injected at an angle of 125° could be deflected and transmitted through the end-cap electrode of the trap and the quadrupole by applying a large negative DC potential to the bottom end-cap electrode.

Control over the kinetic energy of the injected ions was achieved in the BE/quistor/quadrupole instrument described by Schlunegger and co-workers.[10] An electrostatic deceleration lens assembly[36] allowed the injection of the ions from a 3 keV ion beam into the RF ion trap with kinetic energies as low as 5 eV. Ions were injected axially into the trap. A collision gas was admitted directly to the gas-tight trap having only two 1.5 mm holes for ion entry and exit. A maximum of the stored ion signal intensity was observed at a He target gas pressure reading of 10^{-2} Pa, and an injection energy of 20 to 25 eV. At this pressure, the transmitted ion current was reduced to 10 to 20% of its value when no collision gas was added. The authors were able to observe both injected precursor ions as well as their fragments produced during the injection process. No estimate was made of the trapping efficiency, but it appears to have been low.

The coupling of a commercial quadrupole ion trap to a BE mass spectrometer has been described by Cooks and co-workers.[11] In their instrument, the kinetic energy of the injected ions also could be controlled by an electrostatic deceleration stage. The injection energy was varied between 0 and 150 eV. The ions were injected axially into a reduced-size ion trap with a ring electrode radius of 0.5 cm. The He collision gas was introduced directly into the ion trap to a pressure reading of 6×10^{-5} torr outside the trap. The trapped ions were analyzed with the mass-selective instability scan mode of the ion trap. Spectra of injected and trapped ions were reported with S/N ratios exceeding 750. Extensive fragmentation of the injected ions was observed even at injection energies as low as 10 eV. For example, the fragment ions of m/z 219 from perfluorotributylamine could not be injected intact and only a small percentage of the molecular ions of n-heptane survived injection. The presence of both SID and collision-activated dissociation (CAD) fragments indicated that at least a fraction of the injected ions hit an electrode surface. Further evidence for surface collisions was the observed production of hydrocarbon ions in the trap. By variation of the injection energy, energy-resolved daughter spectra were obtained.

Figure 9 shows a daughter ion mass spectrum obtained on this instrument upon injection. The BE section of the instrument was used to select the $[M + H]^+$ ion (m/z 556) of leucine-enkephalin, ionized by fast-atom bombardment (FAB), under moderate resolution of 1500 for injection into the trap. The substantial fragmentation observed at 50 eV injection energy yielded sequence information as is demonstrated in the spectrum in which the sequence ions are labeled using the established nomenclature.

C. Mass Selection by Quadrupole

In order to avoid floating of the RF ion trap at the high voltage potential of the ion source of a sector instrument, both Schlunegger and co-workers as well as Cooks and co-workers constructed instruments in which the mass of the injected ions was selected by a quadrupole mass filter.

An ion trap detector with a reduced-size ion trap (ring radius 0.5 cm) was coupled to a quadrupole mass filter by Cooks and co-workers.[53] The ions were injected axially with kinetic energies between 0 and 150 eV. The He collision gas was introduced to an indicated pressure of 1.7×10^{-4} torr, corresponding to an estimated pressure of 10^{-3} torr within the trap. The trapped ions were analyzed with the mass-selective instability scan mode

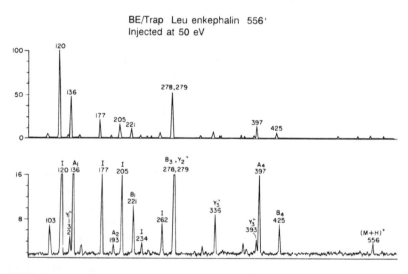

FIGURE 9

Daughter ion mass spectrum of protonated Leu-enkephalin, m/z 556, 50 eV injection energy. The spectrum acquisition time was 5 s. The lower spectrum shows an expanded vertical scale. (From Reference 11; reproduced by permission of Elsevier Science Publishers B.V.)

of the ion trap. Good sensitivity was reported, exemplified by the detection of m/z 92 of toluene from a 40 ppb toluene–water solution leaked into the system through a sheet membrane probe. The efficiency of injecting the ion beam into the trap was measured to be in the range of 10%, but no account was made of the percentage of ions ultimately being stored in the ion trap. The principal findings were consistent with the observations made with the BE–trap hybrid instrument.[11] The production of hydrocarbons through desorption upon collision of the injected ions with the ion trap electrodes was confirmed by experiments with injected Xe^+ ions. The molecular ion of pyrene could be injected intact at injection energies as high as 50 eV. The extensive fragmentation observed at an injection energy of 110 eV was seen as a further indication of surface collisions. When a supercritical fluid chromatography interface was coupled to this instrument, a sample of 80 fmol of anthracene could be detected at a S/N ratio of 20.[54]

Schlunegger and co-workers[38] constructed a quadrupole, quistor, quadrupole instrument in order to study selected-ion/selected-molecule reactions. To exclude interference of the sample neutrals introduced into the ion source with the reactions in the RF ion trap, the respective vacuum chambers were isolated by a pressure differential exceeding 30,000. Ions were mass selected by the first quadrupole mass filter and injected axially into the trap with kinetic energies on the order of 3 eV. He gas was admitted directly to the gas-tight trap up to a pressure of 6×10^{-3} Pa measured outside the trap. With specially designed injection optics, an ion beam injection efficiency of 10% was achieved at an injection energy of 3 eV. It was reported that ions were trapped in significant numbers only when their kinetic energy was below 10 eV. From the injection of O_2^+ ions, it was observed that the trapped ions are more reactive initially, and are then cooled during a cooling period of about 10 ms. The cooled ions were observed to react at the thermal reaction rate constant observed on non-Paul trap instruments. Further aspects of this instrument are discussed in the next section.

VI. QUADRUPOLE, QUISTOR, QUADRUPOLE INSTRUMENT

A. Description of the Instrument

In the following subsections, the operational characteristics of injection and trapping with the quadrupole, quistor, quadrupole instrument at Bern are described in some detail. The instrument has been reported in the literature.[38] Therefore, only a short description is given here. The details needed for the discussion of various aspects of ion injection are given in the respective subsections. The instrument consists of an ion source, a first quadrupole, a quistor, and a second quadrupole (Figure 10). Each of these parts is housed in a separate vacuum chamber and pumped independently by a turbomolecular pump.

The whole ion optics are mounted collinearly, except for the electron beam and the secondary electron multiplier, which are at 90° to the instrument axis. The vacuum chambers are joined together, leaving only an aperture of 3 mm^2 for ion transfer. The path of the ions is completely isolated and shielded from the vacuum housings, so that the ion optics can be floated to any desired potential up to about 500 V.

B. Description of the Ion Optics

The primary goal in designing the injection optics was to find a solution that allows injection of the ion beam into the quistor with high transmission at low kinetic energy. The ions are brought through a hole in the entrance end-cap electrode, which should be as small as possible so that the trapping field in the quistor is only minimally disturbed. Focusing a low energy beam through such a small hole is virtually impossible because of the large injection and diversion angles imposed by the Lagrange-Helmholtz equation (Equation 1).

The solution to this problem was to focus the ion beam at its nominal energy through an orifice ("cone" in Figure 11) placed in front of, but very close to, the quistor entrance end-cap and then decelerate the ions

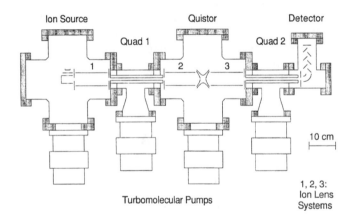

FIGURE 10

Schematic diagram of the vacuum system and ion optics of the quadrupole, quistor, quadrupole tandem mass spectrometer. (From Reference 38; reproduced by permission of John Wiley & Sons, Ltd.)

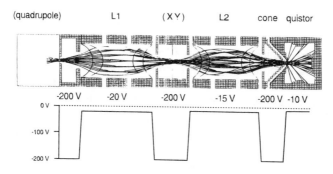

FIGURE 11

Plot of ion trajectories obtained from a SIMION calculation for the injection optics together with potential diagram for positively charged ions. Equipotential contour lines shown are at −30 and −60 V.

over the very short distance between the orifice and the entrance end-cap. By such an ion optical element, the ion beam is made strongly diverging but, due to the short deceleration distance, only after having entered the quistor.

This arrangement for the injection optics was calculated using the SIMION program.[55] In these trajectory calculations, the effects of positional, angular, and energy distributions of the ions in an ion beam were studied. Representative ion beam profiles could be obtained by choosing the initial positions and the initial directions of velocity from within an ellipse in phase space at two different initial kinetic energies. During these calculations, it was found that the SIMION program fails when the ions are decelerated to very low kinetic energies. For certain initial conditions, ions can surmount pure electrostatic potential barriers even if their total initial energy is less than that of the barrier height. Nevertheless, the program is very well suited to calculate the behavior of an ion beam under the influence of focusing and deflecting fields generated by an electrode configuration of arbitrary geometry.

A plot of a SIMION calculation optimized for maximum transmission into the quistor for the specific electrode geometry chosen is shown in Figure 11. The ion beam emerging from the quadrupole (shown at the left side of Figure 11, although the quadrupole rods are not shown) is focused by two lens electrodes (L1 and L2) to the orifice of the cone electrode and the quistor entrance end-cap (at right). Deflector electrodes (placed at "X Y") allow the ion beam to be centered on the orifice. The parameters for this calculation were as follows: phase space ellipse half axis $\Delta\rho_0 = 0.9$ mm and $\alpha_0 = 15°$, initial kinetic energies of 200 and 197 eV, corresponding to final kinetic energies of the injected ions of 10 and 7 eV, respectively.

The first quadrupole for mass selection of the ions to be injected changes the characteristics of the ion beam considerably. From a cylindrically symmetric ion beam injected into a quadrupole field, the mass-selected ions leave the quadrupole field with a cross-shaped beam profile with the lobes pointing toward the quadrupole rods.[56] Because the ions oscillate in the quadrupole field, the exit angles vary widely. Due to the RF field, the width of the kinetic energy distribution of the beam is increased. In addition, a beam entering the quadrupole field at continuous constant intensity leaves the quadrupole field modulated in time with the RF; i.e., the transmission is phase dependent.[56]

Instead of taking all these effects into account separately in the design of the ion injection optics, a simple design was looked for with an acceptance to a broad range of initial conditions. In order to determine realistic initial conditions, the exit geometry and operating parameters of the quadrupole must be considered. The quadrupoles used have a field radius of 3.45 mm.[37] The rods are encased in a cylinder with custom-made entrance and exit apertures, each having central holes of 1 mm radius. The hole size was chosen so as to maximize the ratio of ion/neutral transmission (see Section VI.G). The cylinder with the apertures can be operated at a potential which differs from that of the field axis, which is defined as the mean potential of the rods. For optimum resolution, the field axis is set to a potential corresponding to a nominal ion kinetic energy of 5 to 10 eV, whereas cylinder and apertures are held at a potential which would correspond to an ion kinetic energy of 150 to 200 eV (see Figure 6). Thus, the fringing fields at the end of the quadrupole are traversed quickly by the ions and transmission in and out of the quadrupole is maximized. Thus, realistic initial conditions for the simulation appeared to be a radial spread on the order of the exit aperture radius, a large angular spread corresponding to a maximum ion exit angle of about 15°, and an energy spread of several electronvolts.

Because the ionization time for the ionization methods used (EI and CI) in combination with quadrupole mass selection is much longer than the quadrupole RF period, the phase will be averaged over the injection period. However, if a pulsed method of ionization were to be used in combination with a quadrupole mass filter, the ionization pulse should be synchronized to the RF phase for optimum transmission. A phase dependence has been observed for the pulsed extraction of ions from the quistor into the second quadrupole, as well as for the coupling of two RF devices with almost, but not exactly, the same frequency (see Section VI.H).

With these injection optics, the ion beam is sprayed into the quistor, and it can be expected that shortly after injection the ions will be spread throughout the quistor volume, a situation similar to that pertaining shortly after *in situ* ionization.

C. Injection Efficiency

The efficiency of the optics for ion injection into the quistor was measured as a function of ion kinetic energy with a floating electrometer. Figure 12 shows the current measured on the exit end-cap electrode of the quistor, including the current through the extraction hole. The transmission into the quistor is almost constant for injection energies down to about 20 eV and then drops approximately linearly to zero for zero injection energy. It is interesting to compare the value at 10 eV to the injection efficiency derived from the SIMION calculations (Figure 11): The measured injection efficiency is 30% and the simulated efficiency 56%, which is quite reasonable agreement considering the rather crude field approximations in the injection hole vicinity used for the calculations. The agreement may be even better if the current into the ring electrode was included. A ring electrode current is expected for slow ions (cf. Figure 11).

D. Trapping Process

In the quadrupole, quistor, quadrupole instrument ions are decelerated electrostatically from 200 eV to a few electronvolts immediately before entering the quistor fields. Trapping is achieved by collisional relaxation of the ions in the quistor with the buffer gas helium, at a pressure of 0.18 Pa. Typical injection currents are a few times 10^{-10} A for an injection pulse duration of 10 ms. Usually, the ion source and the mass-selecting quadrupole are first set to yield the desired ion, after which the injection optics voltages are optimized. The number of trapped ions is

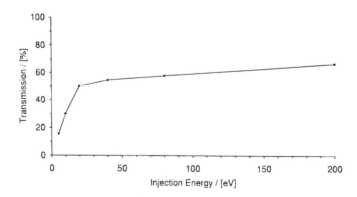

FIGURE 12

Ion current onto a closed extraction end-cap during injection as a function of nominal ion kinetic energy.

then dependent on the injection energy, the quistor RF amplitude, and the buffer gas pressure within the quistor.

At a He buffer gas pressure of 0.18 Pa, the trapped/extracted ion signal intensity of N_2^+ is maximized for a q_z value of 0.44, an injection energy of 5 eV, and an extraction voltage of 30 V. The variation of the signal intensity as a function of buffer gas pressure, injection energy, and working point, keeping all other parameters fixed, is shown in Figures 13 to 15, respectively. Under all conditions tested, increased pressure (Figure 13) always improved the signal intensity up to a limit of about 4 Pa, where ion extraction begins to break down because of collisions occurring during extraction. Even with no He added to the quistor, a very small trapped ion signal is observed. This observation is attributed to trapping by col-

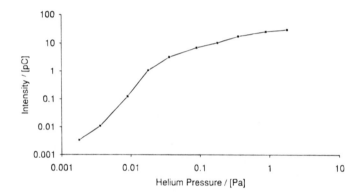

FIGURE 13
Trapped ion signal intensity as a function of He buffer gas pressure.

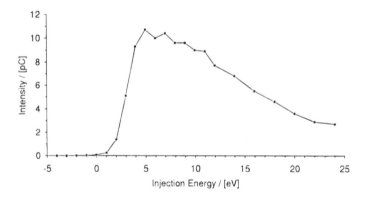

FIGURE 14
Trapped ion signal intensity as a function of injection energy.

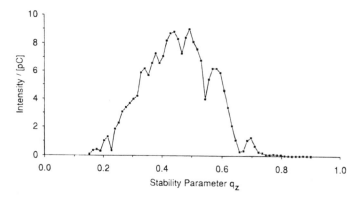

FIGURE 15
Trapped ion signal intensity as a function of quistor RF voltage amplitude.

lisions with background molecules. The background pressure is on the order of 4×10^{-7} Pa. As soon as helium is added at a partial pressure of $>10^{-3}$ Pa, the signal intensity is increased dramatically. The steepest increase in ion signal intensity is proportional to the third power of He pressure. From these results it can be concluded that under these conditions of ion/neutral mass ratio, ion energy, and stability parameter, at least three collisions are needed to trap the ions. At pressures above 2×10^{-2} Pa the signal intensity, while increasing, begins to flatten out as does the multiple collision probability at higher pressures.

The variation in injection energy (Figure 14) reveals that the signal intensity is increased continuously as the injection energy is reduced to values at which the ions do not enter the quistor at all, and the signal drops to zero. The trapping process itself is by far more efficient at low injection energies than is indicated by the signal increase in Figure 14 alone. The strong decrease in injection transmission below 20 eV as shown in Figure 12 is more than compensated for by the much more efficient trapping.

Typical data for variation of the trapping field amplitude are presented in Figure 15. They show that only a restricted range (q_z between 0.2 and 0.7) of the stability region exists (extending from $q_z = 0$ to $q_z = 0.908$), over which ions can be injected and trapped in significant numbers. At low q_z values, no signals were detected at all, in agreement with the threshold observed by other groups (Figure 7), whereas when $q_z = 0.908$ was approached, the ion signal again became very weak but still detectable. As can be seen from Figure 15, strong variations in signal intensity exist within the range of trapping, and are reproducible when the other parameters are held constant. These variations are attributed to resonance effects occurring upon injection or extraction or both.[30,31,50]

Thus far, only one parameter at a time has been varied (Figure 13 to 15). However, it was found that the ion signal intensity lost upon variation of one of injection energy, trapping voltage, or extraction voltage could be recovered partially by adjusting the other two parameters. For example, Figure 16 displays the ion signal intensity as a function of injection energy when the trapping and extraction voltages were adjusted for maximum signal for each value of injection energy. Upon comparing Figure 16 with Figure 14, in which the trapping and extraction voltages were not adjusted, surprisingly it is found that even at 50 eV injection energy, almost 50% of the optimum signal at 5 eV can be obtained. At very low energies, the signal that can be regained by adjusting the trapping and extraction voltages is even more pronounced. The optimum values of the trapping voltage and the extraction voltage used to obtain the data of Figure 16 are given in Figure 17 and 18, respectively. Although some scattering is found in the data, general trends can be seen in both Figures 17 and 18: the higher the injection energy, the higher the optimum value for the trapping voltage (which is given as the stability parameter q_z in Figure 17), and in turn, the higher the extraction voltage required (Figure 18). Of course, the optimum trapping voltage must lie within the boundaries for stable trapping, as is reached with the flat maximum in Figure 17 at injection energies above 80 eV.

E.   Relaxation Process

As mentioned in Section VI.B, the injected ions are spread over almost the entire quistor volume after the first collisions that cause the ions to be trapped. This assumption is supported by the observation of only

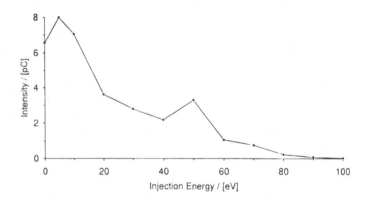

FIGURE 16
Trapped ion signal intensity as a function of injection energy, with trapping voltage and extraction voltage adjusted for maximum signal at each injection energy.

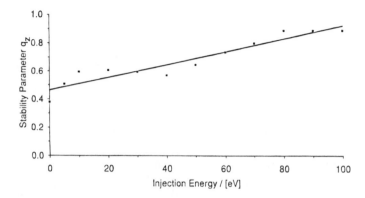

FIGURE 17
Optimum stability parameter as a function of injection energy.

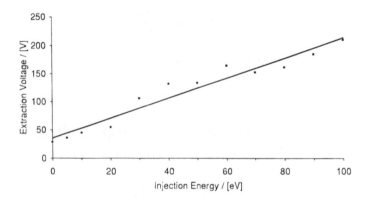

FIGURE 18
Optimum extraction voltage as a function of injection energy.

a very weak extraction signal shortly after injection (Figure 19). Only the ions that are close to the axis when the extraction pulse is applied can be extracted. Further collisions with the buffer gas atoms dampen the ionic motion toward the center of the quistor, and many more ions can be extracted successfully and detected, as shown by the strong increase in the extraction signal after a few milliseconds. This sampling effect also has been observed for internal ionization.[57] For long storage times the ion extraction signal is again reduced due to storage losses. The data in Figure 19 were obtained by injecting N_2^+ ions for an injection time of 100 μs.

Macroscopically, the relaxation process can be described by a damping term in the equation of motion (see Volume I, Chapter 2):

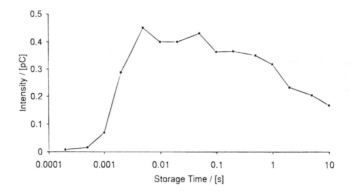

FIGURE 19
Extraction signal intensity as a function of storage delay following injection.

$$m\ddot{x} = -\frac{e}{K}\dot{x} - e\frac{\partial \Phi}{\partial x} \tag{6}$$

where K is the ionic mobility. By the substitution

$$x = e^{\frac{-at}{2}}y \qquad \alpha \equiv \frac{e}{mK} \tag{7}$$

Equation 6 can be brought into the form of the Mathieu differential equation for y with the stability parameters a_y and q_y,

$$a_y = a_0 - \frac{\alpha^2}{\Omega^2} = a_0 - \frac{e^2}{m^2 K^2 \Omega^2} \tag{8}$$

$$q_y = q_0 \tag{9}$$

where a_0 and q_0 are the stability parameters for nondampened motion. When the injected ions are spread initially over a disc of radius r_{inj}, and when those ions within a disc of radius r_{extr} can be extracted, the extracted charge Q for short times t after injection is given by:

$$Q = eN\frac{r_{extr}^2}{r_{inj}^2}e^{\alpha t} \tag{10}$$

Upon fitting the data of Figure 19 to Equation 10, the ion motion damping factor $\alpha = 2736$ s^{-1} and the ratio $r_{inj}/r_{extr} = 10$ can be determined (Figure 20). Using the literature value of 20.9 cm^2 V^{-1}s^{-1} for the standard ionic mobility, K_0, of N$_2^+$ ions in He,[58] the pressure in the quistor is calculated to

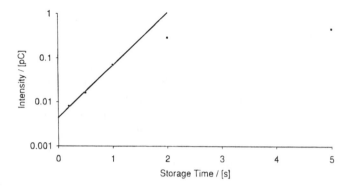

FIGURE 20
Relaxation analysis of the extraction signal.

be 0.18 Pa. This calibration procedure agrees to within a factor of 2 the indirect estimates of the He pressure in the quistor (see Section VI.G).

The stopping distance of an ion with initial velocity v_0 in the absence of an external force is given by

$$b = \frac{v_0}{\alpha} = \frac{m}{e} K v_0 \qquad (11)$$

For an ion with m/z 28, a kinetic energy of 1 eV, and a damping factor of 2736 s^{-1}, the stopping distance is calculated as $b = 96$ cm. This distance is almost 2 orders of magnitude larger than the He target thickness, which is 1.4 cm between the end-cap electrodes. Thus, it would appear that on one hand, injected ions do not have to be stopped completely in order to be trapped, but rather must undergo a relatively small change in momentum. On the other hand, it can be expected that the trapping efficiency could be improved by increasing pressures up to about 10 Pa. Unfortunately, not only detection of the trapped ions but also ion injection fails at those pressures.

F. Trapping Efficiency

In order to obtain an absolute measure for the trapping efficiency, the following experiment was performed. N_2^+ from air was injected into the quistor containing 0.18 Pa of He buffer gas. The instrument was tuned to maximum extraction signal, which was achieved at a nominal injection energy of 2 eV and a quistor RF amplitude corresponding to $q_z = 0.38$. Subsequently, the ion beam current emerging from the first quadrupole was completely deflected from the hole in the entrance end-cap of the quistor onto the cone electrode using the beam-steering deflection plates of the injection optics. The ion current ($I = 2.5 \times 10^{-10}$ A) on the cone electrode was measured with a floating electrometer, leaving constant all of the potentials except those of the steering plates. Then, the beam was again injected into the quistor during the ionization time $\Delta t = 10$ ms. After a relaxation delay of 10 ms the stored ions were extracted and their charge, $Q = 1.7 \times 10^{-14}$ C, was collected with a Faraday collector located directly behind the second quadrupole. The charge was measured using the quadrupole electrometer and signal integration on the digital storage oscilloscope. From these measurements the overall injection, trapping, and detection efficiency was calculated to be $Q/I \, \Delta t = 0.7\%$.

G. Pressure in the Ion Trap

Admission of neutral molecules to the quistor can be controlled independently of the ion source conditions; this is achieved by an excellent pressure differential isolation and separate inlet systems for the two vacuum chambers. The pressure differential is achieved by a three-stage pumping system with separate pumps for ion source, first quadrupole, and quistor, and with apertures of 2 mm diameter in between for gas flow restriction. The pressure differential that can be achieved between the chambers that house the ion source and the quistor is 30,000, when the turbomolecular pumps are running at full speed.

Two separate inlets to the quistor are available for the introduction of buffer and reaction gases. The gases are introduced through leak valves with capillaries leading directly into the quistor. Therefore, the pressure in the quistor is higher than the pressure in the quistor vacuum chamber. For calculations of ion/molecule reaction rate constants, the neutral density inside the quistor, where the reaction takes place, must be known. When the pressure is measured in the quistor vacuum chamber rather than in the quistor itself, it must be corrected by the pressure differential factor between the pressure inside the quistor, p_i, and the pressure outside the quistor p_o, which is given by

$$\frac{p_i}{p_o} = \frac{S}{L} \tag{12}$$

where L is the conductance of the quistor and S is the pumping speed applied to the quistor vacuum chamber. Both L and S depend on the mass of the neutral particles, thus the pressure differential is expected to depend on the mass. The conductance of the quistor is inversely proportional to the square root of the mass. For low molecular weight, the pumping speed of a turbomolecular pump is limited by the rotor speed and decreases only slowly with molecular weight,[57] while for gases of high molecular weight it is limited by the conductance of the pump inlet port and the rotor inlet channels. Therefore, the pressure differential factor (Equation 12) increases somewhat slower than the square root of the mass for compounds of low molecular weight and becomes constant for compounds of high molecular weight. Between methane and high molecular weight compounds, the pressure differential factor is about a factor of 3 different for the turbomolecular pumps used.[60]

H. Synchronization

All RF devices have phase-dependent characteristics; consequently, every timing event that is shorter than a few RF periods is critical. As already pointed out, the transmission of a quadrupole is phase dependent (Section VI.B), thus the phase of the RF drive potential must be considered when a pulse of ions is injected into the quadrupole. Similarly, the trapping efficiency for the injection of a pulse of ions into the quistor and the ion extraction efficiency are phase dependent. For optimum efficiency, the ion pulse must be synchronized to the respective RF phase.

For the quadrupole, quistor, quadrupole instrument, the following synchronization problems were encountered. For the discussion, injection and extraction are considered separately. Because the ion injection time is on the order of milliseconds, which is much longer than the RF period (0.77 μs) of the quistor, the phase dependence of injection into, and trapping in, the quistor would be averaged. However, due to the fact that the RF generators used for quistor and quadrupoles each has a 1.3 MHz quartz oscillator, there is a phase effect during injection. Between these oscillators a small frequency difference occurs, which is on the order of a few Hertz. The small difference between the frequencies of the first quadrupole and the quistor causes the relative phase of the two generators to oscillate with a period on the order of 1 s. Because both quadrupole transmission and quistor injection/trapping efficiency are phase dependent (Sections II.B and VI.B), the number of ions trapped in the quistor during subsequent injection pulses varies with the same period (ca. 1 s). The signal fluctuations caused by this phase oscillation could be eliminated by using only one common frequency source for both RF generators, and the total transmission, injection, and trapping efficiency could be maximized if the relative phase of the two generators can be selected.

It is well known that the extraction efficiency depends strongly on the RF phase at the time when the extraction pulse is applied.[61] For the quadrupole, quistor, quadrupole instrument this phase effect is the most pronounced; therefore, a circuit was built to synchronize the extraction pulse to the quistor RF phase. In addition, there was the same phase problem between the quistor RF and that of the second quadrupole, as discussed above for the first quadrupole and the quistor.

I.   Acquisition

The various stages of an experiment are determined by a pulse sequence (Figure 21). During the ionization pulse, the electron beam is admitted to the ion source so as to generate ions. The desired educt ion, mass selected by the first quadrupole, is injected and accumulated in the quistor. Once the electron beam has been arrested, a time delay allows for the formation of product ions. Once the detection pulse is initiated, the product ions are extracted from the quistor. Those ions having a m/z ratio matching that set for the second quadrupole reach the detector, whereupon a current pulse is generated in the electron multiplier and in the detection electronics. This pulse sequence is repeated for each product ion mass in order to obtain the complete product ion mass spectrum.

Due to the pulsed nature of the ion signal, special detection electronics must be used. A boxcar integrator could be used to integrate the ion current for a short period immediately following a detection pulse, and to hold the signal until the next detection pulse. Because digital data acquisition is used, acquisition gating and signal integration can be performed by the data system. The current pulse is broadened by the electrometer[62] and then digitized at 50 kHz by the analog-to-digital converter of the data system. The pulsed nature of the ion signal permits detection of negative ions as well with a simple modification of the detection circuit (Figure 22). For negative ions, the collector of the electron multiplier must be set at a high positive voltage. A current pulse arriving at the collector can be decoupled from the collector with a high voltage capacitor. The circuit is similar to one used for single ion counting.

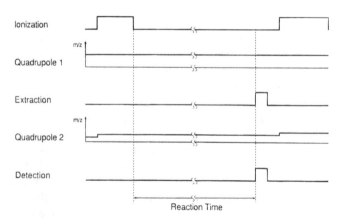

FIGURE 21
Pulse sequence.

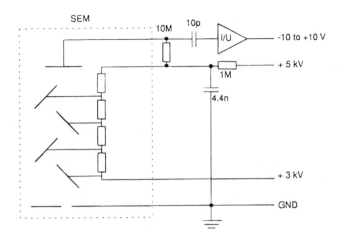

FIGURE 22
Negative ion pulse detection circuit.

The signal integration subroutine of the data acquisition program was programmed so that both positive and negative ions could be detected.

J.   Studies of Ion/Molecule Reactions

Ion/molecule reactions can be studied readily in an RF ion trap because of the long observation times accessible. However, reaction rates determined by RF ion trap experiments must be viewed cautiously, as the effective ion temperature may be considerably different from that of the neutral gas reactant.

The kinetic energy of an ion varies with the phase of the RF trapping field. Therefore, the mean kinetic energy may correspond to a temperature considerably greater than the neutral gas temperature.[3] If the ion collides with a neutral when its kinetic energy is high, part of the collision energy may be converted to internal energy. Because many ion/molecule reactions vary strongly with ion kinetic or internal energy or both, their observed reaction rate constants will depend on RF ion trap parameters.

Several studies have explored effective ion temperatures via ion/molecule reactions with well-known energy dependences. For the Ar^+/N_2 charge exchange reaction, effective ion energies between 0.3 to 0.1 eV were found for pressures of nitrogen in the range 4×10^{-5} to 1×10^{-3} Pa.[63] For the reaction of $O_2^{+\cdot}$ with methane, an effective internal ion energy of 0.12 eV and a kinetic energy of 0.08 eV was found for partial pressures of 5×10^{-3} torr He and 5×10^{-6} torr methane.[63,64]

The reaction of $O_2^{+\cdot}$ with methane has been used for probing effective ion temperatures on the quadrupole, quistor, quadrupole instrument.[38]

In the presence of a large background partial pressure of water, a reaction rate constant close in magnitude to the literature value for room temperature was observed. From this observation it was concluded that the effective ion temperature in the trap is near-thermal after an initial cooling period.

In a subsequent study,[65] the background water was eliminated by careful bake-out of the system and a cold trap for the He buffer gas. It was then found that effective ion temperatures are reduced for lower q_z values (see also Reference 64) and for increased methane pressures (Figure 23). At low pressures of methane, the observed rate constant was almost an order of magnitude greater than the literature value of 5.2×10^{-12} cm^3 mol^{-1} s^{-1} for room temperature.[66] As the methane pressure was increased, the observed rate constant approached the literature value. Moreover, additional evidence was found that the observed rate constant is not dependent exclusively on the methane pressure, but also on the mixing ratio of methane and He. The data presented in Figure 23 were acquired at a constant He pressure of 0.36 Pa and with $q_z = 0.5$. $O_2^{+\cdot}$ ions were generated by electron impact and injected into the trap with a kinetic energy of 3 eV. The methane pressure within the trap was calibrated by determining the rate constant for the reaction of CH_4^+ with methane, which is known to proceed at the temperature-independent rate of 1.13×10^{-9} cm^3 s^{-1} mol^{-1}.[67]

It is known that vibrationally excited O_2^+ ions are efficiently relaxed by collisions with methane, but not with He.[68] Therefore, it can be concluded that thermal conditions only are reached when both ion motion

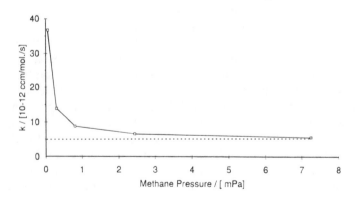

FIGURE 23
Rate constant for the reaction of O_2^+ with methane as a function of methane pressure observed on the quadrupole, quistor, quadrupole instrument. For comparison, the literature value for the room temperature rate constant is also given (dotted line).

and internal energy (from ionization or RF heating via collisions) are efficiently quenched.

VII. CONCLUSIONS

Mass-selected injection of ions has now been demonstrated experimentally by several groups with mass selection carried out both externally and within an RF ion trap. External mass selection is an advantage to the extent that any desired analyzer can be chosen, and that the tuning of the trap for maximum trapping efficiency is independent of the mass-selection tuning. Mass selection within the RF trap during ion injection appears to be adequate for the exclusion of low mass ions.

Although the objective of the initial experiments for mass-selected ion injection was to collect the product ions of high energy collisions in the RF ion trap, the method may find more applications as a means for filling the trap with a specific ion species, which may be studied subsequently within the trap during the ion storage period. By using mass-selected injection, ions of interest though in low abundance may be accumulated in the trap to higher ion densities than that imposed when the space-charge limit is attained by ions in higher abundance.

The choice of the injection energy depends on the desired effect. For fragmentation of the injected ions, a high injection energy (ranging from 10 eV up to some kiloelectronvolts) is required. For trapping injected ions intact, low injection energies (a few electronvolts and less) are essential. The injection of low energy ions requires special injection optics that allow deceleration and high transmission of the ion beam into the trap.

For applications in mass spectrometry, trapping under collisional conditions has been the method of choice. A low molecular weight buffer gas (He, in most cases) has been used for the trapping of ions injected into an RF ion trap. For the injection of ions of high m/z ratio, buffer gas mixtures have been demonstrated to improve trapping efficiency.

Injection of ions from an external source only makes sense when source and trap are pumped differentially. The ion optics for transfer of the ions from the source to the trap must seek to maximize the ratio of ion/neutral transmission.

ACKNOWLEDGMENT

The author extends his grateful thanks to Professor Urs P. Schlunegger for his continued inspiration and support which made possible both this contribution and the underlying research.

REFERENCES

1. Paul, W.; Steinwedel, H. Apparatus for Separating Charged Particles of Different Specific Charges. German Patent 944,900, 1956; U.S. Patent 2,939,952, 1960.
2. Stafford, G.C., Jr.; Kelley, P.E.; Syka, J.E.P.; Reynolds, W.E.; Todd, J.F.J. *Int. J. Mass Spectrom. Ion Processes.* 1984, 60, 85.
3. March, R.E.; Hughes, R.J. *Quadrupole Storage Mass Spectrometry.* Chemical Analysis Series, Vol. 102. Wiley Interscience: New York, 1989.
4. Cooks, R.G.; Glish, G.L.; McLuckey, S.A.; Kaiser, R.E. *Chem. Eng. News.* 1991, 69 (12), 26.
5. March, R.E. *Int. J. Mass Spectrom. Ion Processes.* 1992, 118/119, 71.
6. Louris, J.N.; Cooks, R.G.; Syka, J.E.P.; Kelley, P.E.; Stafford, G.C., Jr.; Todd, J.F.J. *Anal. Chem.* 1987, 59, 1677.
7. Fischer, E. *Z. Phys.* 1959, 156, 1.
8. Ghosh, P.K.; Arora, A.S.; Narayan, L. *Int. J. Mass Spectrom. Ion Phys.* 1977, 23, 237.
9. O, C.-S.; Schuessler, H.A. *J. Appl. Phys.* 1981, 52, 1157.
10. Suter, M.J.-F.; Gfeller, H.; Schlunegger, U.P. *Rapid Commun. Mass Spectrom.* 1989, 3, 62.
11. Schwartz, J.C.; Kaiser, R.E., Jr.; Cooks, R.G.; Savickas, P.J. *Int. J. Mass Spectrom. Ion Processes.* 1990, 98, 209.
12. Curtis, J.E.; Kamar, A.; March, R.E.; Schlunegger, U.P. *Proc. 35th ASMS Conf. Mass Spectrom. Allied Topics.* Denver, 1987; p. 237.
13. Ho, M.; Hughes, R.J.; Kazdan, E.; Matthews, P.J.; Young, A.B.; March, R.E. *Proc. 32nd ASMS Conf. Mass Spectrom. Allied Topics.* San Antonio, TX, 1984; p. 513.
14. Louris, J.N.; Amy, J.W.; Ridley, T.Y.; Cooks, R.G. *Int. J. Mass Spectrom. Ion Processes.* 1989, 88, 97.
15. Dehmelt, H.G.; Walls, F.L. *Phys. Rev. Lett.* 1968, 21, 127.
16. Wuerker, R.F.; Goldenberg, H.M.; Langmuir, R.V. *J. Appl. Phys.* 1959, 30, 441.
17. Dehmelt, H.G. *Adv. Atom. Mol. Phys.* 1967, 3, 53.
18. Wineland, D.J.; Drullinger, R.E.; Walls, F.L. *Phys. Rev. Lett.* 1978, 40, 1639.
19. Vedel, F. *Int. J. Mass Spectrom. Ion Processes.* 1991, 106, 33.
20. Cox, K.A.; Morand, K.L.; Cooks, R.G. *Proc. 40th ASMS Conf. Mass Spectrom. Allied Topics.* Washington, D.C., 1992; p. 1781.
21. Nand Kishore, M.; Ghosh, P.K. *Int. J. Mass Spectrom. Ion Phys.* 1979, 29, 345.
22. Todd, J.F.J.; Freer, D.A.; Waldren, R.M. *Int. J. Mass Spectrom. Ion Phys.* 1980, 36, 371.
23. O, C.-S.; Schuessler, H.A. *Int. J. Mass Spectrom. Ion Phys.* 1981, 40, 53.
24. Alford, J.M.; Williams, P.E.; Trevor, D.J.; Smalley, R.E. *Int. J. Mass Spectrom. Ion Processes.* 1986, 72, 33.
25. Lammert, S.A.; Cooks, R.G. *J. Am. Soc. Mass Spectrom.* 1991, 2, 487.
26. Glish, G.L.; Goeringer, D.E.; Asano, K.G.; McLuckey, S.A. *Int. J. Mass Spectrom. Ion Processes.* 1989, 94, 15.
27. Bonner, R.F.; Fulford, J.E.; March, R.E.; Hamilton, G.F. *Int. J. Mass Spectrom. Ion Phys.* 1977, 24, 255.
28. Gabrielse, G.; Haarsma, L.; Rolston, S.L. *Int. J. Mass Spectrom. Ion Processes.* 1989, 88, 319.
29. Gerlich, D.; Kaefer, G. *Astrophys. J.* 1989, 347, 849.
30. Schwartz, J.C.; Cooks, R.G.; Weber-Grabau, M.; Kelley, P.E. *Proc. 36th ASMS Conf. Mass Spectrom. Allied Topics.* San Francisco, 1988; p. 634.
31. Williams, J.D.; Syka, J.E.P.; Kaiser, R.E., Jr.; Cooks, R.G. *Proc. 38th ASMS Conf. Mass Spectrom. Allied Topics.* Tucson, AZ, 1990; p. 864.
32. McLuckey, S.A.; Glish, G.L.; Asano, K.G. *Anal. Chim. Acta.* 1989, 225, 25.

33. Zworykin, V.K.; Morton, G.A.; Ramberg, E.G.; Hillier, J.; Vance, A.W. *Electron Optics and the Electron Microscope.* John Wiley & Sons: New York, 1957.
34. Vestal, M.L.; Blakley, C.R.; Ryan, P.W.; Futrell, J.H. *Rev. Sci. Instrum.* 1976, 47, 15.
35. Kofel, P.; McMahon, T.B. *Int. J. Mass Spectrom. Ion Processes.* 1990, 98, 1.
36. Suter, M.J.-F.; Stepnowski, R.M.; Schlunegger, U.P. *Rapid Commun. Mass Spectrom.* 1989, 3, 417.
37. QMA 400, BALZERS AG, Balzers, Liechtenstein.
38. Kofel, P.; Reinhard, H.; Schlunegger, U.P. *Org. Mass Spectrom.* 1991, 26, 463.
39. Marshall, A.G.; Schweikhard, L. *Int. J. Mass Spectrom. Ion Processes.* 1992, 118/119, 37.
40. Rasch, J. Dissertation, Frankfurt/M., Germany, 1983.
41. Larsen, B.S.; Wronka, J.; Ridge, D.P. *Int. J. Mass Spectrom. Ion Processes.* 1986, 72, 73.
42. Wang, M.; Marshall, A.G. *Anal. Chem.* 1989, 61, 1288.
43. Rempel, D.L.; Gross, M.L. *J. Am. Soc. Mass Spectrom.* 1992, 3, 590.
44. Hop, C.E.C.A.; McMahon, T.B.; Willett, G.D. *Int. J. Mass Spectrom. Ion Processes.* 1990, 101, 191.
45. McIver, R.T.; Hunter, R.L.; Bowers, W.D. *Int. J. Mass Spectrom. Ion Processes.* 1985, 64, 67.
46. Hunt, D.F.; Shabanowitz, J.; McIver, R.T., Jr.; Hunter, R.L.; Syka, J.E.P. *Anal. Chem.* 1985, 57, 765.
47. McIver, R.T., Jr. *Int. J. Mass Spectrom. Ion Processes.* 1990, 98, 35.
48. Kofel, P.; Allemann, M.; Kellerhals, Hp.; Wanczek, K.-P. *Int. J. Mass Spectrom. Ion Proc.* 1985, 65, 97.
49. Kaiser, R.E., Jr.; Louris, J.N.; Amy, J.W.; Cooks, R.G. *Rapid Commun. Mass Spectrom.* 1989, 3, 225.
50. Pedder, R.E.; Yost, R.A.; Weber-Grabau, M. *Proc. 37th ASMS Conf. Mass Spectrom. and Allied Topics.* Miami Beach, FL, 1989; p. 468.
51. Van Berkel, G.J.; Glish, G.L.; McLuckey, S.A. *Anal. Chem.* 1990, 62, 1284.
52. Bier, M.E.; Hartford, R.E.; Herron, J.R.; Stafford, G.C. *Proc. 39th ASMS Conf. Mass Spectrom. Allied Topics.* Nashville, TN, 1991; p. 538.
53. Morand, K.L.; Horning, S.R.; Cooks, R.G. *Int. J. Mass Spectrom. Ion Processes.* 1991, 105, 13.
54. Pinkston, J.D.; Delaney, T.E.; Morand, K.L.; Cooks, R.G., *Proc. 39th ASMS Conf. Mass Spectrom. Allied Topics,* Nashville, 1991, 160.
55. Dahl, D.A.; Delmore, J.E. *SIMION Version 4.0.* Idaho National Engineering Laboratory: Idaho Falls, 1988.
56. Campana, J.E.; Jurs, P.C. *Int. J. Mass Spectrom. Ion Phys.* 1980, 33, 119.
57. Bonner, R.F.; March, R.E.; Durup, J. *Int. J. Mass Spectrom. Ion Phys.* 1976, 22, 17.
58. Ellis, H.W.; Pai, R.Y.; McDaniel, E.W.; Mason, E.A.; Viehland, L.A. *Atom. Data Nucl. Data Tables.* 1976, 17, 177.
59. BALZERS AG, *Turbo-Molecular Pumps,* PM 800 062 PE (8802), Balzers, Liechtenstein.
60. TPU 240, BALZERS AG, Balzers. Liechtenstein.
61. Dawson, P.H.; Lambert, C. *Int. J. Mass Spectrom. Ion Phys.* 1974, 14, 339.
62. EP 112, BALZERS AG, Balzers, Liechtenstein.
63. Basic, C.; Eyler, J.R.; Yost, R.A., *J. Am. Soc. Mass Spectrom.* 1992, 3, 716.
64. Nourse, B.D.; Kenttämaa, H.I. *J. Phys. Chem.* 1990, 94, 5809.
65. Reinhard, H. *Gasphasenchemie gespeicherter organischer Molekül-Ionen,* Dissertation, Bern, Switzerland, 1992.
66. Adams, N.G.; Smith, D.; Ferguson, E.E. *Int. J. Mass Spectrom. Ion Processes.* 1985, 67, 67.
67. *Gas Phase Ion-Molecule Reaction Rate Constants Through 1986.* Ikezoe, Y.; Matsuoka, S.; Takebe, M.; Viggiano, A. Eds., Ion Reaction Research Group of the Mass Spectroscopy Society of Japan: Tokyo, 1987.
68. Durup-Ferguson, M.; Böhringer, H.; Fahey, D.W.; Fehsenfeld, F.C.; Ferguson, E.E. *J. Chem. Phys.* 1984, 81, 2657.

Chapter 3

ELECTROSPRAY AND THE QUADRUPOLE ION TRAP

Scott A. McLuckey, Gary J. Van Berkel, Gary L. Glish, and Jae C. Schwartz

CONTENTS

I.   Introduction 90

II. Instrumentation 91

III. Singly Charged Ions 95
     A.   Detection Limits 97
     B.   Structurally Informative Fragmentation 99
          1.   Collision-Induced Dissociation 99
          2.   Interface- and Injection-Induced Fragmentation . 101
          3.   Random Noise Collisional Activation 103

IV. Multiply Charged Biomolecules 105
     A.   Mass Spectrometry of Peptides and Proteins 105
          1.   Low Resolution MS/MS and MS^n of Peptides
               and Proteins 111
          2.   High Resolution MS/MS of Peptides and
               Proteins 116
          3.   Ion/Molecule Reactions of Multiply
               Charged Ions 118
               a.   Fundamental Aspects 119
               b.   Applications 131
     B.   Mass Spectrometry and MS/MS of Oligonucleotides .. 134

V.   Conclusions and Prognosis 139

0-8493-4452-2/95/$0.00+$.50

Acknowledgments 140

References 140

I. INTRODUCTION

The unifying theme of this chapter is the use of the quadrupole ion trap in mass spectrometry (MS). The term "mass spectrometry", of course, refers collectively to a broad range of techniques that generally fall into the categories of ion formation, mass/charge (m/z) measurement, and ion detection. Each technique has its own set of characteristics such that certain ionization method/mass analyzer/detector combinations are highly compatible while others are not. Systems that fall into the latter category are generally not developed for analytical applications unless the potential advantages of such a combination are worth the effort to overcome the incompatibilities. The subject of this chapter, electrospray ionization combined with the quadrupole ion trap, or electrospray/mass spectrometry (ES/MS), is an example of a mass spectrometer combination that, at least a few years ago, did not appear to be a promising match. Furthermore, it was also unclear that the combination would hold any advantages over electrospray combined with the quadrupole mass filter. Presently, it is increasingly recognized that the quadrupole ion trap is a remarkably powerful tool when coupled with electrospray. The purpose of this chapter is to discuss how and why electrospray is coupled with the ion trap, drawing from both published and unpublished results. These data are intended to illustrate conventional experiments as well as those unique to ion trapping instruments.

The recent impressive developments in ion trap mass spectrometry (ITMS) qualify as some of the most significant advances in the area of mass analysis of the past decade. Overlapping with these developments was the advent of electrospray, a truly major development in the area of ionization. The evolution and characteristics of the still-rapidly developing spray ionization techniques have recently been reviewed.[1-6] Electrospray provides a means for forming gas phase ions from a wide variety of analyte species present in solution. These species typically include polar compounds and ions in solution. Ions in solutions used for electrospray are typically formed via acid-base chemistry but can also arise from charge transfer chemistry[7] or electrochemistry.[8] Perhaps the most remarkable characteristic of electrospray is its tendency to form multiply charged ions from large molecules such as synthetic polymers[9] and biopolymers.[10] Clearly, there are many important analytical applications for species that form singly charged ions via electrospray (see below), but it is the large, multiply charged ions that pose altogether new challenges and opportunities to mass spectrometrists in deriving structural information.

Many characteristics of the quadrupole ion trap can make it a high performance mass spectrometer. These characteristics are discussed throughout this chapter in varying levels of detail. They include *inter alia* mechanical simplicity, small size, MS/MS[11] and MS[n] capabilities,[12,13] wide *m/z* range,[14,15] and high *m/z* resolution.[16,17] These and other aspects of ITMS are described in this chapter as they pertain to analysis of ions derived from electrospray.

The discussion begins with a description of the instrumentation used to couple electrospray with the ion trap. This section is followed with a description of the electrospray and ion trap mass analysis of analyte species that tend to form singly charged ions. Porphyrins, which represent a prototypical class of organic compounds, are used to illustrate what analytical performance might be expected for a wide variety of moderately sized polyatomic species. Heavy emphasis is then placed on biopolymers, the most important application area for electrospray to date. Many of the strengths of ITMS are particularly valuable for the analysis of multiply charged ions and these are illustrated mainly with peptides and proteins. A section on negatively charged oligonucleotides is also included, providing an example of the potential for improved application of MS in this area. Much of the discussion of multiply charged ions involves techniques and capabilities already known to be useful, such as high resolution and MS/MS involving dissociation. However, ion trapping instruments are also well suited to the study of ion/molecule reactions. There is currently interest in pursuing ion/molecule reactions of multiply charged ions both from fundamental and applied points of view. The first studies of ion/molecule reactions of multiply charged biopolymers were performed with a quadrupole ion trap, and several such studies involving peptides and proteins are described. Finally, the current state of the art in ES/ITMS is summarized and a brief prognosis for the combination is given.

II. INSTRUMENTATION

At the time of writing, there are still relatively few electrospray/ion trap systems. To our knowledge, all have employed the "stretched" geometry ion trap (discussed elsewhere in this book) that is provided with commercial ion trap systems. In all cases, ions have been injected into the ion trap axially. Of those described in the literature, one was based on a modified Saturn II system marketed by Varian,[18] while the remainder have been modified versions of the Finnigan Ion Trap Mass Spectrometer (ITMS)[TM].[19-22] These systems share many features, however, the most significant differences appear in the electrospray/vacuum interfaces. The Finnigan electrospray/ion trap instrument was operated with a fundamental frequency applied to the ring electrode of 880.3 kHz, whereas the fundamental frequency applied to all others has been 1.1 MHz. All of the data presented in this chapter were acquired either with the systems at Oak Ridge National Laboratory or with the system at Finnigan MAT. These systems are therefore described in some detail. However, other significantly different means for coupling electrospray with an ion trap can be imagined and may be realized as more workers take an interest in this combination.

The first experiments involving electrospray with a quadrupole ion trap were performed with an ITMS system which had been modified for atmospheric sampling glow discharge ionization.[23] The hardware for the glow discharge source[24,25] was used as the interface, although it was not designed for transporting ions formed in atmosphere into the high vacuum region. Nevertheless, with the addition of two lenses in the glow discharge source, this hardware was used for the initial work with electrospray.[19] This interface is described here, although some minor modifications of the interface have been made more recently.

Figure 1 shows a cross-sectional view of the instrumental configuration used for the data acquired at Oak Ridge (not drawn to scale). A homemade electrospray apparatus is situated before the interface. A dome-tipped stainless steel needle, 5 cm long and 120 μm i.d., is supplied with the analyte-containing solution at a rate of 1 to 5 μl/min. The needle is held at a potential of 3 to 4 kV, with the polarity depending upon the polarity of the ion of interest. The interface consists of two inline apertures, A1 and A2, of diameters 100 and 800 μm, respectively. The plate containing A1 separates atmosphere from a region held at a pressure of roughly 0.3 torr. The plate containing aperture A2 separates the region at intermediate pressure (0.3 torr) from the vacuum system containing the ion trap. The base pressure in the vacuum system is roughly 2×10^{-5} torr. Helium is added to the vacuum system to a pressure of roughly 1 mtorr as measured by a capacitance manometer. Two plate lenses are situated within the interface and facilitate ion transport through this region. The plate containing A1, which is electrically isolated from the vacuum system by a Viton O-ring and nylon screws, is typically held at a potential of 80 to 180 V of the same polarity as that of the electrospray needle. The lenses are each held at the same polarity and at a voltage similar to A1 for gentle ion transport. A potential difference of a few tens of volts between the plate containing A1 and the lenses can result in significant collisional activation and fragmentation in the interface. This phenomenon is highly analogous to the fragmentation induced between the nozzle and skimmer of the more commonly used electrospray interfaces.[26] Ions that issue from A2 are focused and gated into the ion trap with a three-ele-

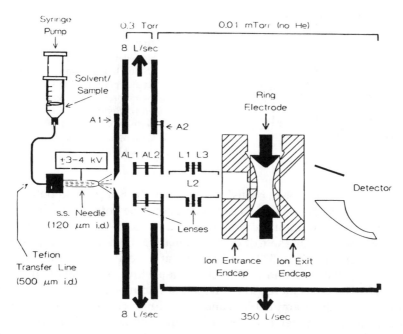

FIGURE 1

Side-view schematic of the ES/ion trap system used in the studies performed at Oak Ridge National Laboratory. (Reprinted with permission from Ref. 19. Copyright 1990, American Chemical Society.)

ment lens system. During ion injection, ions are focused onto the entrance aperture of the entrance end-cap electrode with appropriate voltages applied to the lenses L1, L2, and L3. L2 is comprised of two semicircular plates. During ion injection, both plates of this lens element are held at the same potential. At all other times, the two half-plates are held such that a potential difference of roughly 300 V deflects the ions and prevents them from entering the ion trap. In all experiments described in this chapter, ions were detected by ion ejection into an external detector via resonance ejection. In the case of the Oak Ridge experiments, a modified Galileo model 4873 conversion dynode/electron multiplier detector was used. This detector was modified by electrically isolating the guard-ring and moving the dynode so that it was roughly parallel with the direction of the ions issuing from the ion trap. The guard-ring was held at roughly 1500 V, with the polarity opposite to that of the ions.

A schematic diagram of the Finnigan system is shown in Figure 2. The electrospray source was supplied by Analytica of Branford and consists of a drying chamber in which the grounded electrospray needle is situated, and a heated countercurrent flow of air or nitrogen is utilized. Ions are transmitted through a 500 (μm i.d., 0.635 cm o.d. × 11.4 cm glass

ELECTROSPRAY ION TRAP

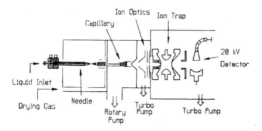

FIGURE 2
Side-view schematic of the ES/ion trap system used in the studies performed in the research laboratories of Finnigan MAT Corp. (Reprinted with permission from Ref. 40. Copyright 1992, John Wiley & Sons, Ltd.)

capillary which has been coated with metal (silver or gold) at both ends so that appropriate potentials can be applied. The entrance to the capillary and the surrounding chamber is held at –3 to –4 kV potential (for positive ions), while the exit of the capillary is held at a small positive potential of some 30 to 300 V. A tube lens at the exit of the capillary aids in focusing the ions, and is held also at a positive potential of some 30 to 300 V. The ions emerge into a region of intermediate pressure (1 to 5 torr) and are then sampled into another vacuum chamber (1 to 5 mtorr) with a nozzle-skimmer arrangement (1-mm skimmer, 3 to 4 mm from the end of the capillary). The ions that pass through the skimmer are focused into a region of lower pressure (1 to 5×10^{-5} torr) and finally onto the entrance aperture of the ion trap during ion injection by a cylindrical gate lens. At all other times, the gate lens is held at a potential sufficiently high (300 V) to prevent ions from entering the ion trap. Ions are detected with a continuous dynode electron multiplier fitted with an off-axis 20 kV conversion dynode.[27]

All ITMS, MS/MS, and MS^n experiments require a programmed set of events typically referred to as the scan function, which, at a minimum, involves an ion injection period and a period in which ions are ejected from the ion trap and into an external detector in a mass-selective fashion. A variety of other intermediate steps may be included depending upon the complexity of the experiment. These steps may be simply time delays to allow for ion/molecule reactions, to allow ions to relax to the center of the ion trap, or to allow ions to desolvate, or they can be mass-selection steps, or collisional activation steps. Ion isolation may be effected in a variety of ways including combined radiofrequency/direct current (RF/DC) applied to the ring electrode and various forms of resonance ejection.[28] In the case of many of the experiments with proteins in the Oak Ridge instrument, ions were subjected to a desolvation step consisting of a rapid scan of the ring electrode RF amplitude in conjunction with supplementary oscillating potential applied to the end-cap electrodes. The purpose of this step is to bring the ions into resonance with the oscillating potential applied to the end-cap electrodes so as to activate the solvated ions collisionally. A rapid scan minimizes the power that the ions absorb so that they are not ejected from the ion trap. This procedure was found to be necessary for some of the Oak Ridge studies because no drying chamber is used in the electrospray region.

III. SINGLY CHARGED IONS

As stated above, the major focus of ES/MS has been the analysis of high molecular weight biopolymers. However, many applications have demonstrated that ES/MS is suited for the analysis of a variety of medium and small molecular weight species, both organic and inorganic. Examples include metal salts,[29,30] ammonium and phosphonium salts,[31] small peptides,[31,32] sulfonated azo dyes,[32] steroids,[32] and transition-metal complexes.[33] As a rule of thumb, if an analyte when dissolved in solution is ionic or can be ionized in solution via acid/base chemistry (i.e., by adjustment of solution pH), it will be amenable to analysis by ES/MS. In contrast to biopolymers, however, small organic molecules typically carry only one or two charges due to fewer ionizable groups and to the close relative proximity of the charges. Nevertheless, the m/z ratios for these smaller molecules are usually within the range (m/z = 10 to 650) accessible under normal operation of the ion trap or are accessible using resonance ejection. Therefore, ion trap performance with electrospray ions derived from small molecules is comparable to performance with ions formed in situ via electron or chemical ionization (EI, CI). Ion trap performance for the analysis of singly charged ions generated by electrospray is illustrated below using porphyrins, an important and ubiquitous class of compounds, for which we have considerable analysis experience using the ES/ITMS combination.[34]

The electrospray mass spectrum of freebase etioporphyrin-III in Figure 3 is representative of the type of spectra obtained from free-base alkyl-substituted porphyrins. Only one major peak is observed in the spectrum corresponding to the protonated molecule, (M + H)⁺. Although significant signal intensity for freebase porphyrins of this type is observed when they are sprayed from a variety of solvents including toluene, toluene:methanol, methylene chloride, and methylene chloride:methanol solutions, addition of acid (in particular, trifluoroacetic up to about 0.5% by volume) to these solvent systems generally provides an enhanced and more stable signal. This signal enhancement upon protonation in solution under acidic conditions is possible because the pyrrolic nitrogens in the macrocycle are basic. The importance of ionization of the porphyrins in solution for optimum ES/MS results is demonstrated by the electrospray mass spectra of an equimolar mixture of six biological porphyrins

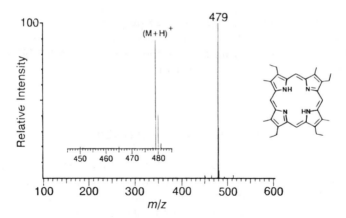

FIGURE 3

The ES mass spectrum of freebase etioporphyrin-III (mol wt 479) obtained by continuous infusion at 5 µL/min. The porphyrin was dissolved in and sprayed from a solvent composed of methylene chloride:methanol:trifluoroacetic acid (10:90:0.1% v:v:v).

shown in Figure 4. The spectrum in Figure 4a was obtained using toluene:methanol:acetic acid (60:40:0.2 percent v:v:v) as the electrospray solvent and the spectrum in Figure 4b was obtained using toluene:methanol:trifluoroacetic acid (60:40:0.2% v:v:v). In both spectra, the porphyrins are observed as the protonated molecules with no fragmentation. Although each porphyrin is present in the mixture at the same concentration, the abundance of the diester component (m/z 595) is enhanced relative to the other components when using acetic acid. The abundances of the porphyrins in the spectrum in Figure 4b are, however, more nearly equal, reflecting the equal concentrations of each in solution. These observations can be rationalized in terms of the acid strength of acetic acid and trifluoroacetic acid used in the electrospray solvent, and in terms of the basicities of each of the six porphyrins. Each addition to the macrocycle of an electron-withdrawing substituent, methyl acetate and methyl propionate in this case, reduces the basicity of the pyrrolic nitrogens. Within this mixture, mesoporphyrin-IX dimethylester is the most basic of the six prophyrins (pK_a = 5.8), coproporphyrin-I tetramethylester is slightly less basic (pK_a = 5.5) and, based on this substitution effect, uroporphyrin-I octamethylester is the least basic. The use of trifluoroacetic (pK_a <1.0), which is a stronger acid than glacial acetic acid (pK_a = 4.74), is necessary to protonate the weakest bases among the porphyrins in this mixture. Because the porphyrins must compete for charge with other basic species in solution, including the other porphyrins, the concentration and strength of the acid used in the electrospray solvent must be sufficient to protonate the least basic of the analytes or it will not be observed.

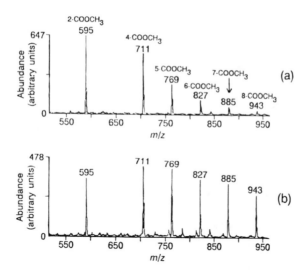

FIGURE 4

The ES mass spectra of an equimolar mixture (10 ± 1 pmol μL^{-1}) of six porphyrin esters obtained by continuous infusion using two different solvent systems. (a) The ES mass spectrum obtained using toluene:methanol:acetic acid (60:40:0.2%, v:v:v) as the ES solvent. (b) The ES mass spectrum obtained using toluene:methanol:trifluoroacetic acid (60:40:0.2%, v:v:v) as the ES solvent. The six porphyrin esters are uroporphyrin-I octamethylester (8-$COOCH_3$, mol wt 942), heptacarboxylporphyrin-I heptamethylester (7-$COOCH_3$, mol wt 884), hexacarboxylporphyrin-I hexamethylester (6-$COOCH_3$, mol wt 826), pentacarboxylporphyrin-I pentamethylester (5-$COOCH_3$, mol wt 768), coproporphyrin-I tetramethylester (4-$COOCH_3$, mol wt 710), and mesoporphyrin-IX dimethylester (2-$COOCH_3$, mol wt 594). (Reprinted with permission from Ref. 34. Copyright 1991, American Chemical Society.)

A. Detection Limits

Detection limits were investigated using flow injection at a few microliters per minute (through a 100-μm i.d. capillary) and optimum solution conditions for condensed phase ionization of the porphyrins. In these experiments, known quantities of porphyrin were injected into the electrospray solvent stream with the analyte eluting from the electrospray needle as a peak as shown by the extracted ion current profiles in Figure 5. In this example, two injections of different quantities of freebase octaethylporphyrin were made as well as a sample blank. As the data demonstrate, <400 fmol of octaethylporphyrin injected produces a spectrum with excellent signal-to-background ratio from which the molecular weight of the porphyrin can be determined. Although the porphyrin is detected at the 18 fmol level, sample impurities introduced via sample handling comprise the major portion of the peaks observed in the spectra, thereby inhibiting the analysis of unknowns. One might expect that analysis of this sample with an online separation at a similar flow rate (for example,

Flow Injection

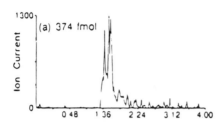

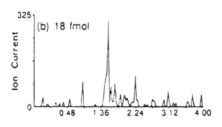

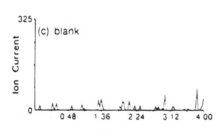

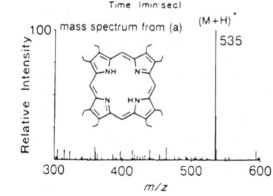

FIGURE 5

The full scan, extracted ion current profiles for m/z 535, the protonated molecule of octaethylporphyrin, obtained during three flow injection analysis experiments in which (a) 374 fmol (900 ms ion injection), (b) 18 fmol (1.5 s ion injection) of analyte were injected, and (c) a blank was injected (1.5 s ion injection). The octaethylporphyrin samples and the blank, both in toluene:methanol:acetic acid (60:40:0.2%, v:v:v), were injected (0.5 µL) into a ES solvent stream of the same solvent composition. The ES mass spectrum at the bottom of the figure was obtained at the peak maximum in the extracted ion current profile of Figure 7a. (Reprinted with permission from Ref. 34. Copyright 1991, American Chemical Society.)

packed-capillary high performance liquid chromatography HPLC, at 1 μL/min) would eliminate much of the chemical noise resulting in electrospray mass spectra with good signal-to-background. In any case, these levels of detection are similar to those found with large biopolymers under conditions optimized for their analysis on this instrumentation.[19] The lowest levels of detection, regardless of the analyte, are to be expected with flow injection or an online separation using the smallest diameter capillaries and columns at flow rates at or below 1 μL/min, thereby minimizing volumetric dilution of the analyte by the solvent.

B. Structurally Informative Fragmentation

The lack of fragmentation in electrospray ionization is advantageous for molecular weight determination and mixture analysis, but limits the structural information obtained. The ES/ITMS combination presents several means, some of which are unique to the ion trap, to obtain structurally informative fragment ions from singly charged ions generated by electrospray. The methods discussed here are single-frequency collision-induced dissociation (CID), electrospray interface fragmentation, fragmentation induced upon injection of externally generated ions into the trap, and random noise CID.

1. Collision-Induced Dissociation

Collision-induced dissociation in the ion trap generally proceeds via multiple low-energy collisions of the ions with the helium bath gas. These collisions are promoted by subjecting the ions to kinetic excitation via a supplementary RF voltage applied to the end-cap electrodes, the frequency of which matches one of those of the periodic motion of the ions. This technique is often referred to as resonant excitation. The energy deposited in the ion can be controlled somewhat by varying the amplitude of the resonant excitation applied to the end-cap electrodes, and by varying the duration of the excitation period. Under typical conditions for performing CID in the ion trap with small, singly charged ions such as the porphyrins, q_z is set equal to approximately 0.2 or greater, and a supplementary RF voltage of 0.5 to 1.0 $V_{(p-p)}$ is applied for 10 to 20 ms. Under these conditions, nearly complete dissociation of the parent ion is affected and the intensity of the product ions formed is maximized.

In general, we have found that porphyrin molecular or pseudomolecular ions formed by electrospray fragment by single-frequency CID in the trap to yield product ions characteristic of the different substituent groups on the macrocycle.[13,34] (This observation is similar to the results from low-energy CID of porphyrins reported on beam instruments.[35])

Additional stages of CID and mass analysis (MS)^n can be performed with the porphyrins to investigate the remaining substituent groups on the macrocycle and to determine the substituent group fragmentation pathways. These points are illustrated by the data in Figure 6, acquired from protonated vanadyl etioporphyrin-III. Shown are the MS/MS, MS^3, MS^4, and MS^5 spectra of the reaction sequence

$$(M + H)^+ \rightarrow (M + H - 29)^+ \rightarrow (M + H - 29 - 15)^+$$
$$\rightarrow (M + H - 29 - 15 - 15)^+ \rightarrow \text{products} \quad \textbf{(1)}$$

For the first several stages of collisional activation, essentially only one dissociation pathway exists when relatively gentle conditions are used to excite the ions. Also, essentially no ion loss is associated with the four stages of collisional activation. Similar product ions also can be obtained using relatively energetic collision conditions in the ion trap in a single stage of MS/MS. Figure 7 is the MS/MS spectrum of protonated vanadyl

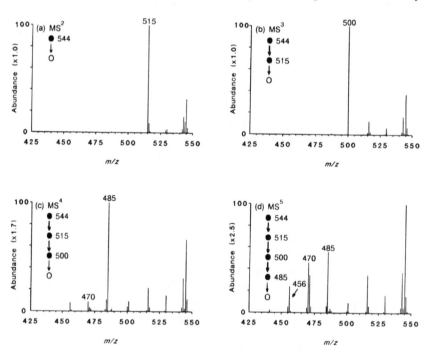

FIGURE 6

MS^n spectra of vanadyl etioporphyrin-III. The abundance of each spectrum is scaled to the abundance of the initial parent ion, protonated vanadyl etioporphyrin-III, m/z 544. After the MS/MS experiment (a) the next generation parent ion was not isolated, i.e., the higher mass ions in (b) to (d) remain in the ion trap from the earlier steps of the experiment. (Reprinted with permission from Ref. 13. Copyright 1991, Elsevier Science Publishers.)

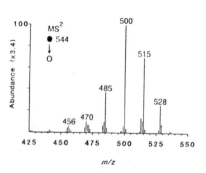

FIGURE 7

MS/MS spectrum of protonated vanadyl etioporphyrin-III acquired with a tickle voltage = 10 V. (Reprinted with permission from Ref. 13. Copyright 1991, Elsevier Science Publishers.)

etioporphyrin-III obtained with a very high amplitude tickle voltage (10 $V_{(p-p)}$) applied to the end-cap electrodes. From comparison of the data in Figures 6 and 7, it can be seen that the degree of dissociation that can be achieved using the MS^n sequence is similar to, or more extensive than, that of MS/MS using high amplitude kinetic excitation. Advantages of the MS^n approach are diminished loss of ions and product ion genealogical information (parent → first generation product → second generation product → etc.).

2. Interface- and Injection-Induced Fragmentation

As with other electrospray interfaces,[36] the voltages on the lenses in our atmosphere/vacuum interface (i.e., voltages on A1, AL1, and AL2 in Figure 1) can be adjusted so that the molecular species generated by electrospray can be fragmented through energetic ion/molecule collisions in the interface region during transport to the ion trap. The electrospray mass spectrum of vanadyl etioporphyrin-III, shown in Figure 8a, was obtained at an injection RF voltage amplitude corresponding to a low-mass cutoff of m/z 50, with the interface lenses adjusted to minimize fragmentation. The spectrum in Figure 8b was obtained at this same low-mass cutoff, but the voltages on the interface lenses were increased. The increased voltage gradient in the interface results in more energetic ion/molecule collisions leading to multiple cleavages of the peripheral alkyl substituents on the molecule. Note that more fragmentation is observed in this spectrum than is observed in the single-frequency CID MS/MS spectrum of the protonated molecule, even with a high amplitude tickle voltage (Figures 6a and 7).

Fragmentation can also be induced upon injection of ions from the interface into the ion trap by adjustment of the RF voltage amplitude (low-mass cutoff) applied to the ring electrode during ion injection. The spectrum in Figure 8c was obtained under focusing conditions established to minimize fragmentation in the interface, but at a higher low-mass cutoff (m/z 150). By injecting porphyrin ions into the trap at higher ampli-

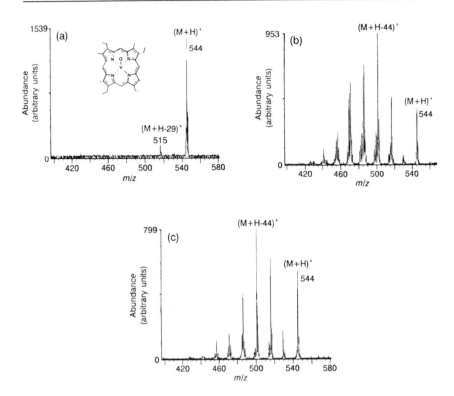

FIGURE 8

The ES mass spectra of vanadyl etioporphyrin-III demonstrating control over injection-induced fragmentation by varying the lens voltages in the ES interface and by varying the amplitude of the fundamental RF voltage on the ring-electrode during ion injection. (a) ES mass spectrum obtained at an injection RF amplitude corresponding to a low mass cutoff of m/z 50 with the interface lenses adjusted to minimize fragmentation (A1 = 90 V, AL1 = 30 V, and AL2 = 90 V). (b) ES mass spectrum obtained at an injection RF amplitude corresponding to a low mass cutoff of m/z 50 with the interface lenses adjusted to maximize fragmentation (A1 = 190 V, AL1 = 170, and AL2 = 200 V). (c) ES mass spectrum obtained at an injection RF voltage amplitude corresponding to a low mass cutoff of m/z 150 with the interface lenses set at the same voltages as in (a). (Reprinted with permission from Ref. 34. Copyright 1991, American Chemical Society.)

tudes of the fundamental RF voltage on the ring electrode, a similar degree of fragmentation to that observed in the spectrum in Figure 8b can be induced.

Neither interface fragmentation nor injection-induced fragmentation is mass selective; i.e., for a particular set of operating parameters, analytes with a relatively wide range of molecular weights can be fragmented. As such, these fragmentation techniques are of utility both in the analysis of pure compounds by continuous infusion and in combination with an online separation, particularly for the analysis of unknowns. By way of contrast, a practical limitation in obtaining structurally informative fragmentation on the ion trap using the normal single frequency method of CID is imposed by the mass-selective nature of the process. The m/z ratio of an ion must be known in advance so that the parameters affecting CID can be optimized (for example, q_z, resonance excitation amplitude, time, and frequency). When an online separation involving compounds of unknown mass is performed, CID cannot be preoptimized and it is not possible at present to optimize single-frequency CID on the chromatographic time scale. The compounded tuning requirements in MS^n experiments, which are often needed to obtain additional structural information, makes acquisition of such data online highly impractical.

An added bonus of interface fragmentation and injection-induced fragmentation is that the variable parameters (lens voltages and low-mass cutoff) used to control fragmentation in these two techniques can be controlled automatically during data acquisition using the ITMS™ software. When these fragmentation methods are used with an online separation, this automation ability allows both mass spectra that contain only molecular species and mass spectra that contain molecular species and fragment ions to be acquired in alternate scans. Thus, both molecular weight and structural information can be obtained in one analysis for each component as it elutes.[34]

3. Random Noise Collisional Activation

A nonmass-selective means of dissociation of trapped ions employing the application of random noise to the ion trap end-cap electrodes was demonstrated recently by our group.[37] This technique is a simple and inexpensive approach to overcoming the single-frequency collisional activation drawbacks of frequency tuning, space-charge effects, limited fragmentation, and the inability to activate parent ions over a range of m/z values.

The relaxed tuning requirements associated with random noise collisional activation makes this dissociation technique promising for MS/MS in online separation applications for which the limits of single-frequency collisional activation are most severe. An example of on-line microbore HPLC/ES/MS/MS using random noise collisional activation is shown in Figure 9.[38] In this experiment, two injections of the porphyrin hematin were made, the first with the random noise signal off, the second injection with the noise signal on. The resulting ion current profiles and mass spectra demonstrate acquisition, requiring two separate hematin injections, of the hematin mass spectrum which is devoid of fragment ions, and the structurally informative product ion spectrum provided via random noise collisional activation. As with interface and injection-induced fragmentation described above, random noise MS/MS product ion spec-

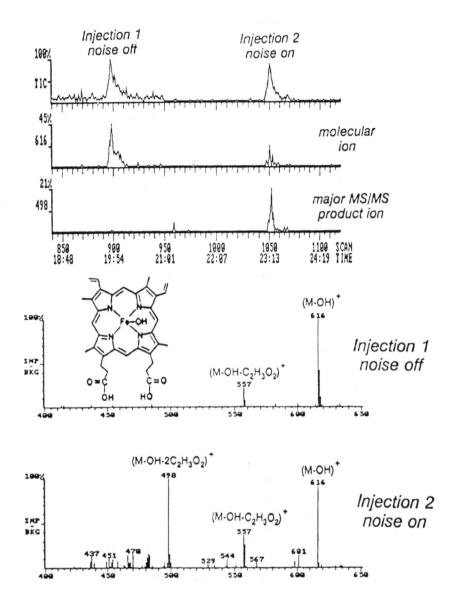

FIGURE 9

On-line microbore HPLC/ES/MS/MS using random noise to effect CID. Two injections of hematin (221 pmol) were made, the first with the noise signal off and the second with the noise signal on. Shown are the total ion current profile, the extracted ion current profiles for the molecular species and major product ion, as well as the averaged mass spectra obtained in both experiments.[38]

tra can be obtained for alternate mass scans under computer control, allowing for the acquisition of molecular weight and structural information in a single experiment.

IV. MULTIPLY CHARGED BIOMOLECULES

Among the most dramatic results to come from electrospray have been those obtained with polymeric species, notably biopolymers. Electrospray provides means both for ionizing and volatilizing high mass biomolecules, the latter being historically a major stumbling block in biological MS. Particularly intriguing is the tendency for the formation of multiply charged ions from electrospray of biopolymers. These multiply charged ions pose new opportunities and challenges to MS in general as well as to the ion trap specifically. In considering the ion trap as an analyzer for multiply charged biomolecules, one must address its figures of merit relative to those of other types of m/z analyzers. At first glance, the limited dynamic range of the ion trap relative to beam-type instruments stands out as its fundamental weakness and might lead one to conclude that the ion trap is a poor choice as a mass analyzer for electrosprayed ions. Indeed, the limited ion storage capacity of the ion trap is currently an issue in its use as an analytical mass spectrometer. For this reason, a number of measures have been employed to mitigate the effects of limited ion storage. However, as is demonstrated here and elsewhere in this chapter, the potential of the ion trap as a high performance mass analyzer for high mass ions coupled with its relative low cost clearly justify efforts to overcome problems associated with limited storage capacity and ion-ion interactions. This section is devoted to illustrating the present capabilities of the ion trap for providing molecular weight and structural information from multiply charged biopolymers, as well as a description of the ion/molecule reactions of such species. The results presented herein are intended both to indicate what can be done now and the areas where improvements are possible and desirable.

A. Mass Spectrometry of Peptides and Proteins

A principal advantage of ES/MS in the area of biotechnology is the accuracy with which molecular mass can be determined relative to other approaches. As first demonstrated with the quadrupole mass filter, mass accuracies of 0.005% or better constitutes an improvement of two to three orders of magnitude over other non-MS techniques.[2] The inherent limitations to mass accuracy with the quadrupole ion trap have not been well quantified. Most experiments involving the m/z analysis of biomolecules employ resonance ejection. Resonance ejection is essentially a frequency measurement. Mass accuracy is therefore dependent upon the accuracy with which ion frequency can be determined and the precision with which ion frequencies are reproduced. Factors that affect ion frequency other than m/z are the amplitude and frequency of the drive signal, the ion trap dimensions, and any other fields that may arise due to electrode imperfections or other ions. Furthermore, the amplitude and stability of the resonance ejection signal as well as the pressure of the damping gas affect the number of cycles required for ion ejection and, hence, the time at which they exit the ion trap. The relative importance of these parameters and the quantitative dependence of mass accuracy on each have yet to be studied in detail. Cooks et al. (see Chapter 1, this volume) have recently reported mass accuracies of 0.005% using peak matching with singly charged biomolecules formed by cesium ion bombardment. However, the work performed thus far with electrospray and the ion trap has not focused on mass accuracy. Most of the work reported to date has demonstrated mass accuracy to be in the range of 0.2 to 0.5%. The wide range in mass accuracy is due to the wide range of scan speed and resonance ejection power used to acquire the data as well as wide variations in numbers of ions, background pressures, etc. A major limitation in much of the reported data is due simply to the fact that the commercial system, which was originally designed for gas chromatography/mass spectrometry (GC/MS) applications, assigns six data points per m/z unit. When the mass range is extended by resonance ejection, as it usually is for ES/MS experiments, even fewer data points are available to define peak position. The data presented in this chapter, therefore, do not reflect the kind of mass accuracy that may be expected from an ion trap system designed for high mass biomolecules. Nevertheless, until the issue of ion trap mass accuracy is fully addressed, it should be regarded as a weak point relative to other mass analyzers, including the quadrupole mass filter, sector instruments, and Fourier transform-ion cyclotron resonance (FT-ICR) spectrometers.

Mass accuracy with the commercial ion trap is maximized when no mass range extension is employed which requires that the ions of interest must fall within the nominal m/z range of m/z 10 to 650. The mass scale provided by the commercial data system can be calibrated only by EI of the mass calibration compound, perfluorotributylamine. Therefore, a good scenario for mass measurement with present-day equipment would involve no m/z range extension and a mass-scale calibrated by EI of perfluorotributylamine. A poor scenario is illustrated with the help of Figure 10; which shows the ES/MS of horse skeletal muscle myoglobin acquired with the Oak Ridge instrument using a m/z range extension factor of 7. The m/z scale in this case was calibrated using the ES/MS of cytochrome c, also acquired with a m/z range extension factor of 7. The mass resolution reflected in this spectrum is between 500 and 1000, while the

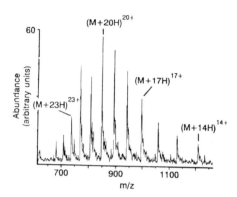

FIGURE 10
ES mass spectrum of horse skeletal muscle myoglobin acquired with a 7-fold extension of the m/z range over that of the standard ion trap. (Reprinted with permission from Ref. 19. Copyright 1990, American Chemical Society.)

mass accuracy is about 0.15 percent. Increasing the m/z scale by a factor of 7 while maintaining the scan rate of the RF amplitude applied to the ring electrode effectively increases the mass scan rate to over 38,000 m/z units per second. Given that the data system assigns a data point every 28 µs and that (roughly) only 25 µs are allotted per m/z unit, it is clear that the digitizing rate is limiting the assignment of peak position. As demonstrated below and elsewhere in this book (see, for example, Chapter 1), mass resolution is inversely related to scan rate so that both mass resolution and mass accuracy are degraded by extending the mass range without a compensating reduction in scan speed and increase in digitizing rate.

From the foregoing discussion, it might be expected that both mass resolution and definition of peak shape could be improved by reducing the scan rate and acquiring more data points across the peak. Figure 11, which shows ES mass spectra of renin substrate, bears this out. The main spectrum was acquired at the "normal" scan rate, whereas the insert zooms in on the $(M + 4H)^{4+}$ ion and shows the result when the scan rate is reduced by a factor of 200. The isotopic distribution of the quadruply charged ion is clear. However, the enhanced resolution at the slower scan speed cannot be accounted for by an increase in the number of data points across the peak alone. Scan rate has a direct effect on the inherent mass resolution of the ion trap and not just on the apparent resolution, as it might have via an effect on the number of data points across the peak. (The effect of scan rate on mass resolution is addressed specifically in Chapter 1.) The effect of scan speed on the appearance of an ES/MS spectrum is illustrated further in Figure 12, which compares several ES mass spectra of angiotensin I in the region of the triply protonated molecule as a function of scan speed and with the theoretical isotope distribution. At $\frac{1}{100}$ the normal scan speed, the mass resolution is roughly an order of magnitude greater than that obtained at the normal scan speed (spectrum not shown). The limit to mass resolution with an ion trap has not yet been

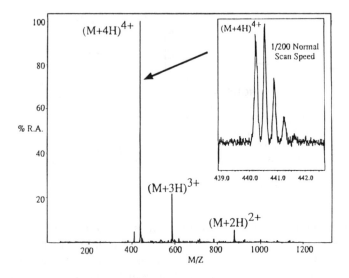

FIGURE 11

ES mass spectrum of renin substrate acquired at the normal scan rate and a narrow scan over the 4+ charge state (inset) acquired at a scan rate reduced by a factor of 200.

established, but we have observed mass resolution for multiply charged biomolecules to exceed 40,000 at scan speeds on the order of a few Daltons per second.

It is also noteworthy that reducing the scan speed also usually results in a reduction in signal-to-noise ratio (S/N). (This effect is reflected in Figure 11 and 12 in the increased noise apparent in the baseline as scan speed is reduced.) This reduction in S/N ratio has been rationalized as being due to the fact that the detector measures ion current which is reduced as the rate at which the ions are ejected from the ion trap is reduced. In those cases in which the integrated ion signal (the area under the ion current vs. time curve) has been determined, it has been shown that the total number of detected ions is independent of scan speed. Appropriate filtering of the data such that the bandwidth of the detection system matches the achieved resolution for a given scan rate, can gain back some of the S/N ratio. (In other words, a slowly emerging peak may be smoothed more than a quickly emerging peak.) Also affecting the S/N ratio is the fact that for some experimental conditions, the tolerance for observing space-charge effects decreases with scan rate. Thus, to achieve high resolution in these cases requires fewer ions in the trap, resulting in a true loss of signal and, therefore, reduced S/N ratio. In these cases, best results are obtained when only ions of interest are stored, requiring all other ions to be ejected prior to the slow resonance ejection scan.

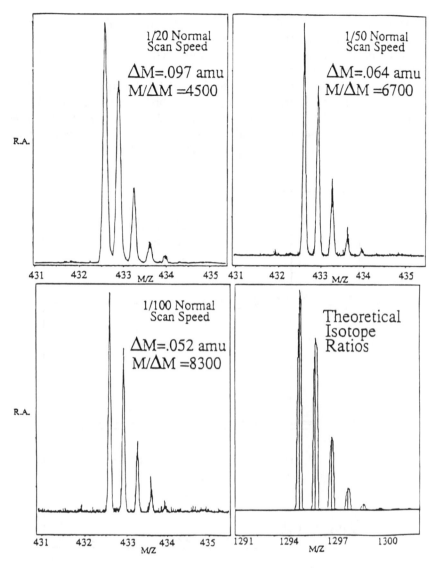

FIGURE 12
ES mass spectra acquired over the region of the 3+ charge state of angiotensin I at various scan speeds along with a simulated spectrum showing the theoretical isotope abundances.

In addition to mass accuracy and mass resolution, the quantity of material required for a molecular mass determination is an important figure of merit. The ion trap, with its well-known advantage in duty cycle relative to beam-type instruments, would be expected to stand out in this regard, provided injection and trapping efficiencies are comparable to transmission efficiencies of the beam-type instruments. As demonstrated above with flow injection analysis of porphyrins, low femtomole quantities of analyte can be measured with the electrospray/ion trap combination for species that are ionized efficiently via electrospray. Low detection limits are also obtainable with the ion trap for electrosprayed peptides and proteins, as is illustrated in Figure 13, which shows both unedited and background-subtracted mass spectra of 8.0 fmol of leu-enkephalin and 7.4 fmol of met-(O)-enkephalin. These spectra were obtained when a portion of a mixture of these species was injected and separated sub-

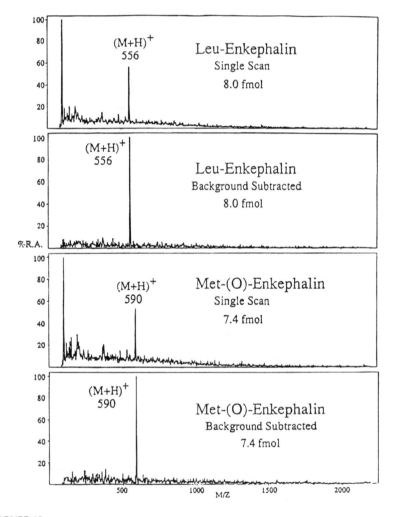

FIGURE 13

ES mass spectra leu-enkephalin and met-(O)-enkephalin following separation by capillary electrophoresis of a mixture containing 8 and 7.4 fmol, respectively.

sequently using capillary electrophoresis, and subjected to online analysis using the electrospray/ion trap combination. This particular example shows results for singly charged ions but larger multiply charged species can also be detected at low femtomole levels.

1. Low Resolution MS/MS and MS^n of Peptides and Proteins

Mass spectrometry/mass spectrometry in the quadrupole ion trap using collisional activation[11] is effected normally by resonance excitation of the parent ions, as discussed in Volume I, Chapter 5. Two of the characteristics associated with collisional activation in the ion trap are the relatively slow nature of ion activation and the high trapping efficiencies for product ions. The slow nature of ion activation accounts for the fact that fragmentation is often restricted to a small number of low critical energy dissociation reactions. (In the case of peptides and proteins, however, the molecule is comprised of many bonds of nearly equivalent strength. Fragmentation, therefore, tends to be rich even under ion trap collisional activation conditions.) The latter characteristic underlies the common observation of high MS/MS efficiencies. Efficiencies sometimes approaching 100 percent are observed when the parent ion population can be excited sufficiently to fragment before a significant number are ejected resonantly from the ion trap. The potential for high MS/MS efficiencies was an important factor in evaluating the ion trap as an analyzer for high mass biomolecule ions. This promise has been borne out for a wide variety of high mass multiply charged biomolecules and constitutes one of the most attractive features of the ion trap for biochemical research. Efficiencies are usually in the tens of percent, especially for the more highly charged species. The following examples are intended to illustrate the strengths and current limitations of the ion trap for determining the primary sequence of peptides and proteins as well as the power of high mass resolution in solving biochemical problems.

Peptides of modest size (500 to 2000 Da) yield rich structural information even with the moderate resolution of the standard ion trap having the upper limit of its m/z range increased to about 1500 m/z units. The MS/MS and MS^n analysis of doubly protonated glu-fibrinopeptide B (M_r = 1569.7, where M_r is the mass of the isotopically most probable molecular ion) is discussed here in some detail to illustrate the point. Figure 14 shows the MS/MS spectrum of the $(M + 2H)^{2+}$ parent ion using a q_z value of 0.3 for excitation of the parent ion. Note the rich fragmentation and

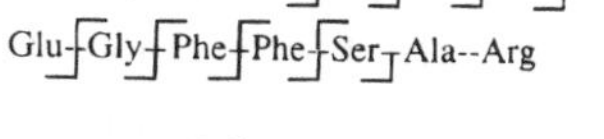

Glu--Gly--Val⌐Asn┤Asp┤Asn┤Glu┐
Glu┤Gly┤Phe┤Phe┤Ser┬Ala--Arg

Scheme 1

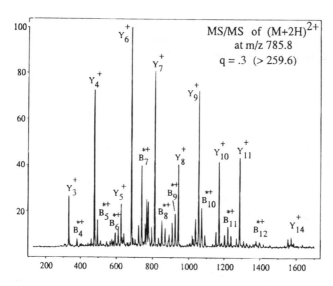

FIGURE 14

Ion trap MS/MS spectrum of the $(M + 2H)^{2+}$ ion derived from glu-fibrinopeptide acquired at a parent ion q_z value of 0.3.

large number of sequence-specific product ions (product ion nomenclature follows that of Roepstorff and Fohlman.[39]) Most of the product ion signal is carried by the complementary B- and Y-type ions and these are the only ions labeled. Scheme I indicates the B- and Y-type ions observed in this spectrum and shows the part of the sequence that they reveal. Lower m/z Y_1^+ and Y_2^+ ions are observed when using a q_z value of 0.1 for parent ion excitation (spectrum not shown). However, at such a low q_z value, high m/z ions are discriminated against, principally the higher m/z Y-type ions.

The mass spectrum of Figure 14 is typical of the ion trap MS/MS spectra of doubly charged peptides of up to 20 residues, in that a large fraction of the peptide sequence can be deduced from the product ions. However, it is a fortuitous situation when the entire sequence is apparent from the MS/MS spectrum of the doubly charged parent ion alone. When the complete sequence is not apparent, a few well-chosen experiments can often furnish the final pieces of information. A variety of experiments are possible, including acquiring the ES mass spectrum under interface conditions established to induce fragmentation (this is not, of course, an MS/MS experiment), the MS/MS analysis of the singly protonated peptide, and MS/MS/MS experiments involving selected product ions. As alluded to above, acquiring MS/MS spectra at several values of q_z can yield more information than acquiring the MS/MS spectrum at a single value of q_z. This situation is due both to different instrumental discrimination effects at different q_z values and possible differences in collisional activation conditions when different well depths are in effect (see Chapter 5). In the case of glu-fibrinopeptide, all of the measures just mentioned have been evaluated and the results are described here.

As discussed briefly earlier, fragmentation induced in the interface can provide useful information when the bulk of the ion signal is due to the analyte alone; i.e., when no mixture of analyte species is found in the electrospray. This situation, of course, prevails when a pure compound is subjected to infusion or flow injection analysis or when analytes are admitted to the electrospray following some form of separation. Fragmentation induced in the interface is a collisional activation experiment under conditions rather different from those in effect within the ion trap. Therefore, the resulting spectrum can be quite different from the ion trap MS/MS spectrum in both the nature of the product ions and in their relative abundances. The parallels and contrasts of ion trap collisional activation vs. interface collisional activation are beyond the scope of this chapter. However, perhaps the most important distinction is that in the ion trap experiment first-generation product ions are not further accelerated in the ion trap experiment. In fact, once formed, product ions are cooled by collisions with the bath gas. On the other hand, a fixed voltage gradient is established in the high pressure interface region that results in multiple collisions and acceleration of all ions. Therefore, first-generation product ions themselves may be accelerated through the high gas density in the interface and undergo CID. For this reason, interface fragmentation using a relatively large voltage gradient tends to be more extensive than that observed in the ion trap MS/MS experiment, with low m/z products tending to be abundant. This collision process is similar to that which occurs in quadrupole collision cells and distinguishes ion trap collisional activation from collisional activation in beam-type instruments. For example, in Figure 14, the presence of large B-type ions distinguishes this spectrum from that taken on a quadrupole instrument. Large B-type ions have been shown to fragment readily under continuous activation. These ions give important information for sequence determination. The interface CID spectrum illustrated in Figure 15 shows the mass spectrum acquired with a potential of 200 V between the capillary and skimmer and a low-mass cutoff of the ion trap during ion injection of m/z 40. Note that several of the lower B- and A-type ions absent in the ion trap MS/MS spectrum are observed here. The major variables in interface fragmentation experiments with the ion trap are the potential gradient and the amplitude of the trapping RF signal applied to the ring electrode. The latter defines the window of ions that can be injected and trapped and, like the acquisition of MS/MS spectra at different q_z values, can be used either to discriminate for or against high or low m/z ions.

Another means for increasing the amount of structural information is to perform collisional activation on first-generation product ions in an

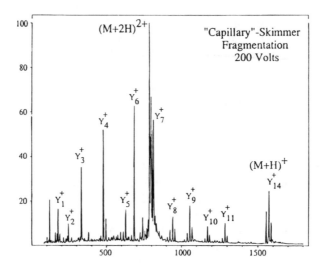

FIGURE 15
ES mass spectrum of glu-fibrinopeptide acquired with a nozzle-skimmer potential difference of 200 V.

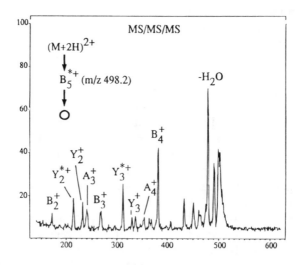

FIGURE 16
The MS/MS/MS spectrum following the sequence $(M + 2H)^{2+} \rightarrow B_5^{*+} \rightarrow ?$ where M = glu-fibrinopeptide.

MS/MS/MS experiment. Figure 16 shows the MS/MS/MS experiment following the sequence $(M + 2H)^{2+} \rightarrow (B_5^{*+}) \rightarrow ?$ where the B_5^{*+} ion is the B_5^{+} ion less a hydroxyl group. Some of the smaller, structurally useful products absent in the spectrum of Figure 14 are produced in this experiment, providing more pieces of the puzzle.

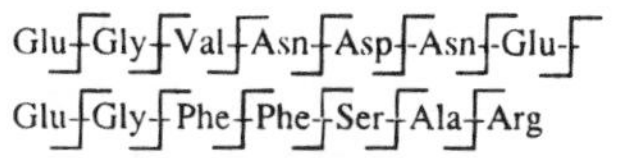

Glu Gly Val Asn Asp Asn Glu
Glu Gly Phe Phe Ser Ala Arg

Scheme 2 (please see additional sizing chart)

of the other major first-generation product ions, such as the Y_4^+ and Y_4^{*+} ions, fill in the blanks for the smaller fragments. Some of the larger fragment ions, such as the Y_{12}^+ and Y_{13}^+ ions, are absent from the MS/MS spectrum of Figure 14, but are observed in MS/MS/MS experiments involving the Y_{14}^+ and Y_{14}^{*+} ions as first-generation products (spectra not shown). With the combined results from the MS/MS spectrum and several MS^3 experiments, all of the cleavages leading to Y- and B-type ions are observed, allowing the entire sequence of the peptide to be determined as indicated in Scheme 2.

The foregoing examples illustrate that there are a variety of tools available with the ion trap coupled with electrospray to determine sequence information of moderately sized peptides. This information is usually obtainable via MS/MS and MS^n studies using conventional ion trap collisional activation. However, single-frequency collisional activation presently requires interactive operator tuning, and it is subject to the problem of parent ion frequency changes due to ion-ion interactions. Both situations are particularly troublesome when online separations are coupled with the system in that little time is available for tuning and, furthermore, optimum tuning conditions may change as the analyte concentration varies as it elutes from the column and through the electrospray needle. Therefore, broad-band kinetic excitation using noise has been studied so as to minimize tuning requirements and to avoid altogether the problem of parent ion frequency shifts due to ion-ion interactions.[37] Figure 17 shows the MS/MS spectrum of neuromedin U-8, an octapeptide, obtained using noise as the resonant excitation signal applied to the end-cap electrodes to effect collisional activation. This spectrum shows all of the product ions apparent in the single-frequency MS/MS spectrum (not shown) with

25%
Rel. Abund.
a2+
y3+
(M+2H)2+
z5+
z6+
(M+H)+
b2+
b7 2+
z3+
b5+
z7+
b7+
200
400
600
800
1000
1200
m/z

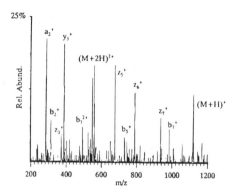

FIGURE 17
The MS/MS spectrum of doubly protonated neuromedin U-8 subjected to collisional activation by random noise. (Reprinted with permission from Ref. 37. Copyright 1992, American Chemical Society.)

some additional product ions at lower abundance due to sequential fragmentation. Only the most abundant ions are labeled, but many more structurally specific products are found in the spectrum. The more extensive fragmentation is due to the fact that product ions are also "in resonance" with the noise signal and can undergo collisional activation. The spectrum of Figure 17 was acquired at some cost in efficiency relative to the single-frequency spectrum (40% vs. 70%) and the spectrum obtained with noise shows some discrimination against low m/z product ions. However, the diminished tuning requirements resulting from the insensitivity to the number of parent ions and the formation of additional product ions make this approach attractive for online separations and ion trap MS/MS.

2. High Resolution MS/MS of Peptides and Proteins

High mass resolution can be effected with the quadrupole ion trap using resonance ejection at reduced scan speeds, as discussed earlier. It is desirable to use high scan speeds provided the information necessary to determine peptide or protein sequence can be obtained with the moderate mass resolution which results. However, instances occur in which high mass resolution is necessary to solve a structural problem. This necessity for high mass resolution becomes increasingly important as the size of the molecule increases and as the number of charges increase. The number of possible combinations of mass and charge that fall within a given m/z window obviously increases with charge state and molecular size.

A particularly challenging aspect of interpreting MS/MS spectra of multiply charged ions is that neither mass nor charge are measured directly, only m/z. Of course, mass is of primary interest when determining sequence so that the possibility for ions of several charge states at a particular m/z ratio makes product ion identification difficult. In cases in which a pair of ions is formed from a single cleavage yielding two observable product ions (i.e. a complementary pair of ions) the ratio of the mass/charge shifts of the product ions from the parent ion indicates the product ion charge ratio. Assuming the parent ion charge to be known, charge assignment is straightforward. However, when one or both of the product ions fragments further or one is not stored, the charge cannot be determined by this simple method. High mass resolution of the product ions, however, can provide charge state information by determination of the spacings between the isotope peaks. Figure 18, the MS/MS spectrum of the $(M + 7H)^{7+}$ ion from human hormone releasing factor (M_r 5039.7), is used to illustrate this point. Note that three peaks are indicated in this figure that could possibly arise from one or more of several different sequence ions. The scan speed required to provide high resolution over the entire mass scale could require many minutes for a single scan. Therefore, high mass resolution is most apt to be analytically useful when small re-

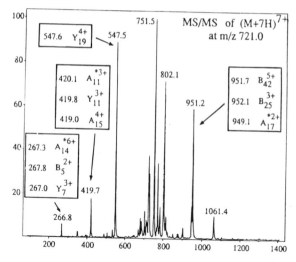

FIGURE 18

The MS/MS spectrum of the $(M + 7H)^{7+}$ ion derived from human growth hormone releasing factor.

gions of the mass scale are of interest. Figure 19 shows the MS/MS spectra obtained by scanning over the m/z regions of the species in question in Figure 18 at $\frac{1}{100}$ the normal scan speed. In all three cases, the spacing between the isotope peaks is 0.33 u, reflecting triply charged ions. As is most often the case, the major product ions are B- and Y-type ions.

All examples of high resolution shown thus far involve the final step of mass analysis. However, instances occur in which high resolution parent ion isolation might be desirable. Such a capability already has been demonstrated with an ion trap[40] and is illustrated here with quadruply protonated renin substrate (M_r 1757.9). Figure 20 shows selected results from the MS/MS analysis of the $(M + 4H)^{4+}$ ion. The upper spectrum shows the results obtained from a low mass resolution parent ion isolation in which all isotopes are present for collisional activation. The portion of the MS/MS spectrum showing the B_9^{2+} product ion m/z region indicates that all of the possible isotope compositions of the product ion are formed by collisional activation. The middle spectrum shows the parent ion region of the spectrum after a slow scan is used to eject the ^{13}C-containing parent ions. The resulting product ion spectrum shows consequently only ^{12}C-containing product ions. The lower spectrum shows the parent ion containing one ^{13}C atom isolated by a series of two slow scans to eject higher and lower m/z parent ions, respectively. The product ion spectrum shows a mixture of ^{12}C-only and one ^{13}C atom-containing product ions in the ratio expected for a parent ion with one ^{13}C in a total of 85 carbon atoms. These results indicate the possibility for parent

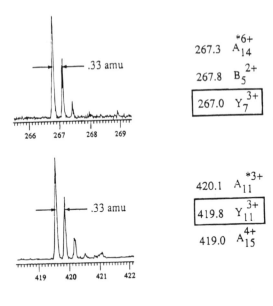

$$267.3 \quad A_{14}^{*6+}$$

$$267.8 \quad B_{5}^{2+}$$

$$\boxed{267.0 \quad Y_{7}^{3+}}$$

$$420.1 \quad A_{11}^{*3+}$$

$$\boxed{419.8 \quad Y_{11}^{3+}}$$

$$419.0 \quad A_{15}^{4+}$$

$$951.7 \quad B_{42}^{5+}$$

$$\boxed{952.1 \quad B_{25}^{3+}}$$

$$949.1 \quad A_{17}^{*2+}$$

FIGURE 19

High resolution MS/MS data for the $(M + 7H)^{7+}$ ion derived from human growth hormone releasing factor demonstrating charge state identification by isotope separation.

ion isolation with greater resolution than is normally performed with the conventional ion isolation techniques, and without significant parent ion loss. The price paid here, however, is in time; i.e., high resolution parent ion isolation can significantly lengthen the scan function. However, in cases in which ionic mixtures require high resolution ion isolation, the ion trap can provide it. This situation can easily arise in MS^n experiments in which first generation product ions are of different mass and different charge but of very similar m/z ratio. In such a case, high resolution ion isolation for the next step of collisional activation may be required.

3. Ion/Molecule Reactions of Multiply Charged Ions

Multiply charged even-electron polyatomic molecules in general, and multiply charged biopolymers in particular, constitute a new class of gas phase ions. The behavior of these species is of interest both from the practical point of view, inasmuch as they can provide structural and molecular mass information, and from a fundamental point of view. Of intrin-

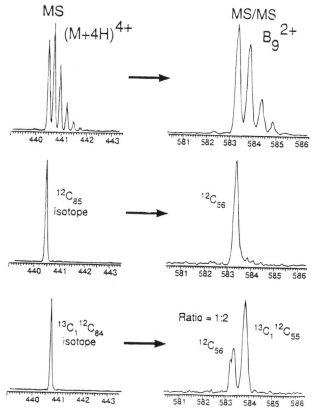

FIGURE 20

High resolution parent ion selection in MS/MS experiments involving ions in the 4+ charge state derived from renin substrate. (Reprinted with permission from Ref. 39. Copyright 1992, John Wiley & Sons, Ltd.)

sic interest is the role of the coulombic field on the uni- and bi-molecular chemistry of these large ions. The effect of the coulombic field on the unimolecular reactions of polyatomic biopolymers is discussed in the section on oligonucleotides. This section emphasizes the bimolecular reactions of multiply charged biopolymers and is illustrated using peptides and proteins.

a. Fundamental Aspects

The study of ion/molecule reactions of multiply charged biopolymers is a new field of research. In fact, initial experiments[41] were carried out in a modified version of the commercial ion trap, and such reports were made as recently as 1990. Since then, reactions of multiply charged ions, such as proton transfer and H/D exchange, have been studied in the gas phase at or near atmospheric pressure[42–45] and in FT instruments.[46]

These types of studies promise to reveal new information about the effects of coulombic repulsion on biomolecular chemistry and, perhaps, on the higher-order structure of multiply charged biopolymers.

Figure 21 illustrates several of the commonly observed phenomena when a family of multiply charged ions derived from the same parent molecule is exposed to a strong gas-phase base in the ion trap. In this figure, the upper mass spectrum shows the ES/MS derived from horse skeletal muscle myoglobin in the absence of any base introduced intentionally into the vacuum system. (Of course, solvent vapors such as water and methanol are always present in the vacuum system; particularly when heated bath gases are not used for desolvation. These vapors themselves may be basic enough to deprotonate the most highly charged species. Indeed, such deprotonation may account for the general observation that the highest charge states for many proteins reported with quadrupole mass filter-based systems are either reduced in relative intensity or absent in the ES/MS acquired with the Oak Ridge ion trap system.) The lower mass spectrum, shown with the same scale on the ordinate, was acquired after dimethylamine was added to the system to a pressure of about 5×10^{-7} torr and after an additional 100 ms reaction period prior

REACTION OF MYOGLOBIN WITH DIMETHYLAMINE

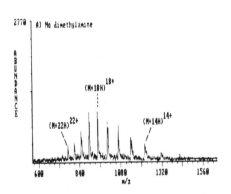

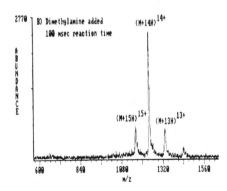

FIGURE 21

ES mass spectra of horse skeletal muscle myoglobin acquired in the absence of dimethylamine (upper) and in the presence of dimethylamine in the vacuum system after a reaction period of 100 ms (lower).

to the mass scan was added. A dramatic shift in charge state distribution occurs resulting, in this case, in a concentration of the charge in the 14+ charge state. As a rule, the higher charge states react faster, with the over-all rate of change in the spectrum decreasing exponentially. Furthermore, as the shift in charge states from high to low decelerates, clustering is often observed. That is, one or more molecules of the base are observed to form proton-bound dimers with one or more of the charge sites on the protein. This clustering can be seen in Figure 21 in the mass shifts asso-ciated with the 15+, 14+, and 13+ charge states, the magnitudes of which correspond to most probable numbers of base molecules of seven to nine.

A more quantitative illustration of the increase in proton transfer rate with increasing charge state is shown in the plot of Figure 22, in which is summarized the rate constant measurements for a series of multiply protonated cytochrome c molecules as a function of charge state. Note that the ordinate scale is linear, and that the highest rate constant mea-sured corresponds to that which might be expected for a fast proton trans-fer reaction for a small protonated molecule (the measured rate constant is nearly 1×10^{-9} cm^3 molecule^{-1} s^{-1}). The 8+ charge state was observed to react too slowly for a reliable rate constant measurement.

It is interesting that the 15+ charge state reacts with a rate constant typical for smaller molecules whose maximum rate constants are described by point charge/dipole and point charge/charge-induced dipole terms.

Multiply protonated cytochrome c/DMA

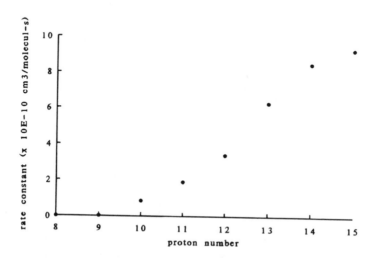

FIGURE 22

A plot summarizing the rate constants measured for proton transfer from cytochrome c to dimethylamine as a function of cytochrome c charge state.

It is likely that this situation is fortuitous. Cytochrome c is observed with higher charge states and these probably react with higher rate constants. However, because dimethylamine is present in the vacuum system throughout the ion accumulation period and ion isolation steps, the higher charge states have been depleted by proton transfer reactions sufficiently to inhibit a reliable measurement of the rate constant. At the other extreme, the rate constants for the lower charge states are difficult to measure, both because long trapping times are required and because there are relatively fewer parent ions at the outset.

The most dramatic examples of clustering have been observed with 1,6-hexanediamine, a molecule with a basic functionality on each end. Figure 23 shows the results when multiply protonated ions derived from bovine insulin are exposed to 1,6-hexanediamine in the ion trap. The upper spectrum shows the ES/MS obtained when no base is intentionally introduced into the vacuum system. The lower spectrum shows the results of roughly 0.5 trapping in the presence of 2×10^{-7} torr of the base. Note that the 5+ ion mostly transfers a proton to the base to give the 4+ ion. A minor degree of clustering is observed with the 5+ ion. The 4+ charge state, on the other hand, shows a small degree of proton transfer to the

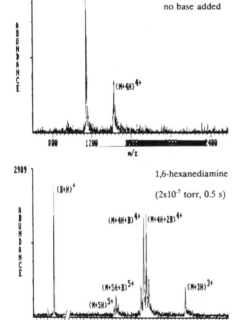

FIGURE 23
ES mass spectra acquired for bovine insulin in the absence of 1,6-diaminohexane (upper) and in the presence of 1,6-diaminohexane (lower) following a reaction period of 0.5 s. (Reprinted with permission from Ref. 49. Copyright 1991, American Chemical Society.)

3+ state but predominantly reacts by clustering. The addition of three molecules of 1,6-hexanediamine is clearly apparent. Unlike the other charge states, the 3+ ion is relatively unreactive both to proton transfer and to clustering. Of course, the 3+ ion has much less time to react in the experiment of Figure 16 because it must be formed first from the higher charge states. However, even at much longer reaction times, it is observed to be relatively unreactive.

The 5+, 4+, and 3+ ions of bovine insulin in reactions with 1,6-hexanediamine illustrate the three limiting cases in interactions involving proton-bound dimers: proton transfer, clustering, and no reaction. This conclusion follows from the kinetic scheme shown below for the overall reaction of the $n+$ ion, MH_n^{n+}, and the base, B:

$$MH_n^{n+} - B] + B \underset{k_b}{\overset{k_c}{\rightleftharpoons}} \left[MH_n^{n+} - B\right]^* \tag{2}$$

$$\left[MH_n^{n+} - B\right]^* \xrightarrow{k_p} MH_{n-1}^{(n-1)+} + BH^+ \tag{3}$$

$$\left[MH_n^{n+} - B\right]^* \xrightarrow{k_s[He]} \left[MH_n^{n+} - B\right] \tag{4}$$

where k_c is the rate constant for collision of the base with a charge site (not necessarily the hard-sphere collision rate of the base with the molecule) to form an ion/molecule complex (Equation 2); k_b is the rate constant for breakup of the complex back to reactants (Equation 2); k_p is the rate constant for breakup of the complex to products which show that proton transfer has occurred (Equation 3); and k_s [He] is the product of the stabilization rate constant and bath gas number density leading to stabilization of the complex to form a cluster ion (Equation 4).

The observed rate constant, k_f, for proton transfer,

$$MH_n^{n+} + B \underset{k_b}{\overset{k_f}{\longrightarrow}} MH_{n-1}^{(n-1)+} + BH^+ \tag{5}$$

is given by

$$k_f = k_c k_p / (k_b + k_p + k_s[He]) \tag{6}$$

Further clustering action is shown in Equation 7, which has the same rate constants for formation and breakup of the cluster complex, $[MH_n^{n+} - 2B]^*$.

$$\left[MH_n^{n+} - B\right] + B \underset{k_b}{\overset{k_c}{\rightleftharpoons}} \left[MH_n^{n+} - 2B\right]^* \tag{7}$$

The observed rate constant, k_{clus}, for the clustering reaction

$$MH_n^{n+} + B + He \xrightarrow{\quad k_{clus} \quad} [MH_n^{n+} - B] + He \qquad (8)$$

is given by

$$k_{clus} = k_c k_s[He]/(k_b + k_p + k_s[He]) \qquad (9)$$

When k_p is much greater than both k_b and $k_s[He]$, k_f approaches k_c and proton transfer is seen to predominate, as in the case of the 5+ charge state. When $k_s[He]$ is much greater than both k_b and k_p, k_{clus} approaches k_c and clustering is seen to predominate, as in the case of the 4+ charge state. When k_b is much greater than both k_p and $k_s[He]$, the complex breaks back up into reactants and no net reaction is seen to occur, as in the case of the 3+ ion.

Of fundamental interest are the underlying reasons *why* the rate constants defined above change in relative magnitude with charge state. This is a very complex question, particularly when one considers that these systems have poorly characterized structures in the gas phase and that their structures almost surely change with charge state. Furthermore, it should be recognized that a number of conformations as well as a number of charge state distributions may contribute to the family of ions at a given charge state. Indeed, intramolecular proton transfer and intramolecular proton-bound dimer formation almost surely take place during the electrospray/desolvation/ion trapping process. It is not known at this time to what extent the ions that are eventually sampled in the ion trap, having been formed there either by collisional activation or by ion/molecule reactions, are structurally pure. Furthermore, the internal energies of these large biomolecules subjected to trapping in the presence of helium bath gas are not yet well characterized, although the evidence is that they are relatively cool (temperature between 300 and 500 K).[47] Without knowing the structures of the ions, the sites of protonation, their internal energies, etc., we can hardly draw firm conclusions concerning the significance of their reactivities. However, we can draw some general conclusions about those reactions that are useful in rationalizing the results that have already been obtained, and are useful in designing new experiments.

If it is assumed that proton transfer proceeds through a relatively long-lived proton-bound dimer complex, a qualitative picture of the energy surface associated with proton transfer and clustering reactions can be drawn by reducing the problem to a single reaction coordinate, so that two-dimensional energy diagrams may apply. The simplest system that exhibits the unique characteristics of a multiply charged biomolecule-ion/molecule reaction is a linear molecule bearing a charge at each end, such as a linear peptide with basic residues at each end. Figure 24 shows

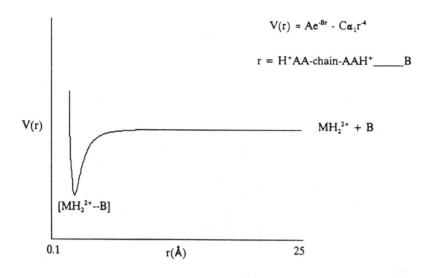

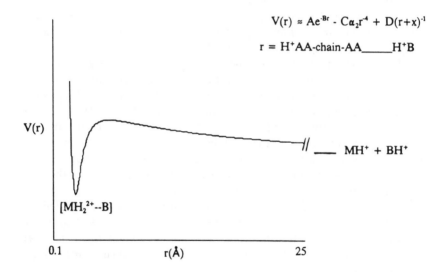

FIGURE 24

Two-dimensional energy diagrams for the ion/molecule reactions of a hypothetical peptide. The upper diagram illustrates an ion/molecule reaction whereas the lower diagram indicates an ion/ion reaction in which the ions are of like charge.

energy diagrams associated with a hypothetical peptide with basic residues at each end of the molecule reacting with a small basic molecule. The upper diagram shows the entrance channel for the reaction in which the neutral base, B, approaches the doubly charged peptide, MH_2^{2+}, along the line connecting the charges. (Details of the potential used for this illustration are not based on known parameters for biomolecules. Indeed, they were chosen only to give features of the diagram deemed to be reasonable in extrapolating from known behavior of proton transfer involving smaller species.) The upper figure includes a repulsive term representing the nuclear repulsion associated with close approach of the proton to the atom of the basic residue to which it attaches. The attractive term is the usual ion/induced dipole term giving the r^{-4} attraction at long distances. Therefore, the entrance channel is simply depicted as a normal ion/molecule reaction with no significant barriers to capture. The lower diagram applies to the exit channel of the reaction, which is the reverse of an ion/ion reaction involving the singly protonated biomolecule, MH^+, and the protonated base, BH^+. This potential contains an additional repulsive term to account for the coulombic repulsion associated with the two protons. This term does not account explicitly for the dielectric constant of the peptide. The D term would be 14.4 eV in the absence of a dielectric constant and would be somewhat less with a dielectric constant. The dielectric constant is expected to be small and, in any case, does not affect the qualitative behavior depicted here.

Charge separation associated with the breakup of the proton-bound dimer into two singly charged products results in a repulsive surface at long range extending over 100 Å, which is overcome by the ion/induced dipole interaction only at short distance. Note that the charge separation aspect to this reaction creates a potential barrier in the energy surface between the minumum energy of the complex and the energy of the products at infinite separation. This feature has important implications for both the bi- and unimolecular reactions[48] of multiply charged ions. The presence of this barrier can provide the ion or complex with kinetic stability even when the products of the decomposition of the ion or complex are more stable thermodynamically. For this reason, some of the multiply charged ions observed from electrospray may not be stable to fragmentation in the thermodynamic sense, but may have long lifetimes in the mass spectrometer due to their kinetic stabilities. It is also for this reason that proton-bound dimer ions may be observable in the ion trap even though they are not stable with respect to dissociation into two charged products (see below).

The energy diagram, presented in Figure 25, shows reaction coordinates for both the entrance channel which proceeds from left to center to form the dimer ion, and the exit channel which proceeds from center to right whereby the dimer ion dissociates to products. In this presentation, the diagram shows the relationships between the respective critical energies associated with both breakup of the dimer ion back to reactants and on to products (the reverse critical energy associated with the charge separation reaction) and the difference between the gas phase basicities of the reactants and products. For proton transfer reactions involving singly

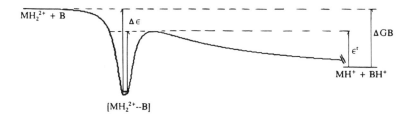

FIGURE 25

A two-dimensional energy diagram of a hypothetical reaction involving peptide M and base B in which, moving left to right, doubly charged M reacts with neutral B to form a cluster ion which then breaks up into singly charged M and singly charged B.

charged ions, reactions generally proceed at the collision rate when the free energy of the reaction is negative. Indeed, the occurrence or nonoccurrence of a proton transfer reaction is often used to bracket gas phase basicities. In the absence of significant reverse critical energies, the difference in the critical energies for breakup of the proton-bound dimer intermediate into reactants or products, $\Delta\varepsilon$, is essentially equal to the difference in the gas phase basicities of the reactants and products. However, as shown in Figure 25, $\Delta\varepsilon$, is not necessarily approximated by ΔGB in the case of proton transfer reactions involving multiply protonated species. The occurrence or nonoccurrence of a proton transfer reaction in the case of a multiply protonated biomolecule is determined by $\Delta\varepsilon$ when the chemical system is under kinetic control, as it is in the ion trap.

When $\Delta\varepsilon$ is large and negative, the reaction proceeds quickly and reactants form products with essentially unit efficiency. When $\Delta\varepsilon$ is large and positive, proton transfer is not observed regardless of the sign of ΔGB. These situations are depicted in Figures 26a and c, respectively. Under some conditions, the proton-bound dimer ion is observed. As indicated above, this occurs when the stabilization rate constant, $k_s[\text{He}]$, exceeds the rate constants for dissociation of the dimer ion, k_p and k_b. This situation is maximized when the respective critical energies are both relatively large and are nearly equal. In these circumstances, the lifetime of the complex is maximized, allowing more time for collisional stabilization. This situation is depicted in Figure 26b. The diagrams of Figure 26, therefore, likely apply to the proton transfer reactions of a series of multiply charged ions from the same molecule with a given base. The strength of the base determines, in part, $\Delta\varepsilon$. Proton transfer proceeds rapidly as long as $\Delta\varepsilon$ remains large and negative. No reactions are observed when $\Delta\varepsilon$ is large and positive, as might be the case for ions with few widely separated charges. When $\Delta\varepsilon$ is small and the number of degrees of freedom is large, as is certainly the case with these systems, the probability for cluster ion formation is maximized.

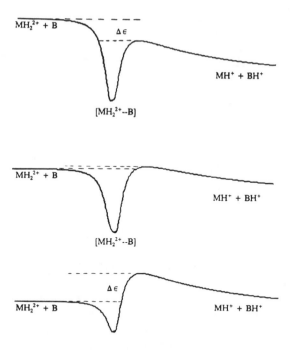

FIGURE 26

Three hypothetical energy diagrams illustrating the three limiting cases in proton transfer reactions involving multiply charged ions.

The extent to which clustering between multiply charged biomolecules and gaseous acids or bases is observed is highly dependent upon reaction conditions. The lifetime of the complex, for example, is determined by the well depth, the number of degrees of freedom in the complex that can share the energy associated with condensation, and the internal temperatures of the reactants. Ion/molecule complex lifetimes are maximized with deep well depths, many active degrees of freedom, and low ion and neutral internal energies. The likelihood for cluster formation is also dependent upon the number density of the bath gas and the relative kinetic energy with which the cluster ion collides with the bath gas; i.e., the bath gas can either "heat" or "cool" the complex depending upon the relative magnitudes of the cluster ion internal energy, the cluster ion kinetic energy, and the temperature of the bath gas. The degree of clustering, therefore, can be expected to vary considerably for reactions occurring at atmospheric pressure vs., for example, those occurring in the rarified atmosphere of the FT mass spectrometer. The ion trap with a bath gas pressure of about 1 mtorr represents an intermediate case.

The fact that clustering is observed to be a dominant reaction channel in the ion trap for some ions was, at first, surprising in light of the general observation that proton-bound dimer formation for smaller molecules is typically not observed to be an efficient process in the ion trap. The inefficiency of small molecule proton-bound dimer formation is due to the fact that the bath gas number density in the ion trap is several orders of magnitude lower than that in a high pressure ion source, wherein proton-bound dimers are readily formed in high abundance. Efficient cluster ion formation for some ions in the ion trap indicates that the inherent cluster ion lifetime is significantly longer than those associated with much smaller cluster ions. Provided some rough estimates can be made for the parameters relevant to cluster ion formation, the inherent lifetimes of the complexes formed in the ion trap can be estimated.

For example, the $(M + 4H)^{4+}$ from bovine insulin rapidly clusters with 1,6-diaminohexane (see Figure 23). If a temperature of 500 K and a hard-sphere collision cross-section of 2000 Å^2 is assumed, the collision rate with helium present at a number density of 3×10^{13} cm^{-3} is roughly 10^5 s^{-1}. Assuming that it takes ten collisions with helium to cool the complex sufficiently for it to remain stable, the stabilization rate, $k_s[\text{He}]$, is about 10^4 s^{-1}. For clustering to occur, the stabilization rate should exceed the sum of the decomposition rates, $k_p + k_b$, by a factor of 10 or so. Therefore, if the assumptions that went into the estimate of $k_s[\text{He}]$ given above are correct, the inherent lifetime of the initially formed complex must be on the order of 1 ms or greater. This is, indeed, a long lifetime relative to those for much smaller systems. The fact that clustering is only a minor process for the $(M + 5H)^{5+}$ ion indicates that the complexes on average have much shorter lifetimes than those involving the $(M + 4H)^{4+}$ ion. For this to be the case, the forward rate constant, k_p, should exceed the sum of the rates for the reverse reaction and stabilization by a factor of 10 or so. As already estimated above, the stabilization rate is about 10^4 s^{-1} so that k_p must be $>10^5$ s^{-1}, which leads, in turn, to an estimated lifetime for complexes that are not observed in the ion trap of <10 µs.

The foregoing discussion provides the background to describe a qualitative picture for the observed kinetic behavior of multiply protonated peptides and proteins exposed to a gaseous base. In its simplest form, the picture can be described qualitatively using Figure 27. This figure shows a hypothetical linear peptide with basic residues present at various points along the chain, and assumes that the basic residues lysine, arginine, and histidine, and the N-terminal amino group are the most likely sites for protonation. The solid horizontal lines represent the $\Delta\varepsilon$ associated with a singly charged proton-bound dimer ion with the proton located at the various sites. The line below and intermediate to the lines for the adjacent lysines is included to indicate the possibility for intramolecular dimer formation that may further stabilize a proton. The dashed horizontal lines represent the hypothetical situation when all of the basic sites are protonated. The coulombic field associated with the

"working model"

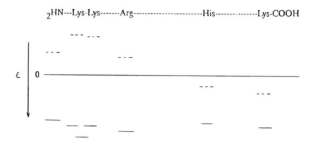

$_2$HN---Lys-Lys-------Arg----------------------His----------Lys-COOH

PA(dimethlyamine) = 223 kcal/mol
PA(1,6-hexanediamine) = 239 kcal/mol
PA(proton sponge) > 239 kcal/mol

FIGURE 27

A simplified scheme depicting the $\Delta\varepsilon$ values for the proton transfer reactions from various sites of a "fully" charged (dashed lines) and singly charged (solid lines) hypothetical peptide to a base. The strength of the base determines the position of the $\Delta\varepsilon = 0$ line.

abundance of like charges significantly alters the $\Delta\varepsilon$ value associated with each protonated site. As indicated above, for a given charge state and ion structure, the point at which $\Delta\varepsilon = 0$ is determined by the strength of the base. Those protons located at sites at which $\Delta\varepsilon$ falls well above the line are therefore vulnerable to scavenging by the base. Those that fall near the line are most likely to form clusters. Those that fall well below the line are relatively unreactive. As each proton is removed by a base all of the other protons are stabilized to an extent dependent upon the relief in coulombic repulsion that they experience.

While this picture clearly is too simple in that it does not account, for example, for intramolecular proton transfer and intramolecular clustering, it does account qualitatively for the observations. Proteins are not simple linear systems and their three-dimensional nature must play a role in the extents to which multiply charged versions are deprotonated by bases of various strengths. Protein folding and solvation of a proton are expected to affect significantly the magnitude of $\Delta\varepsilon$ as the number of protons remaining on the molecule decreases. Nevertheless, the qualitative picture of proton destabilization created by the coulombic field followed by relaxation due to proton transfer to a base appears to be valid for these systems. A quantitative picture, however, cannot be drawn until more is known about the three-dimensional structures of these ions. The use of ion/molecule reactions, however, may provide an indirect means for obtaining such structural information.

b. Applications

Beyond the academic interest in the behavior of multiply charged even-electron ions in the gas phase is the possible information about the structures of biopolymeric ions that these ion/molecule reactions might provide. Indeed, evidence suggests that higher-order structure, native vs. reduced forms of proteins, for example, might be reflected in ion/molecule reactivity.[36] Too little is known at this point to say with much confidence the extent to which these reactions will be useful in studying higher-order structure. However, there are examples in which ion/molecule reactions have been useful in drawing conclusions about the structures of gas phase ions beyond those apparent from CID.

As discussed in the previous section, the tendency for proton transfer or clustering is determined in large measure by the coulombic field experienced by the charges. Therefore, the reactivity of an ion can reveal some indication as to the spacing between charges. In some cases, it could be used along with CID data to draw conclusions about the likely sites of protonation. The behavior of the $(M + 4H)^{4+}$ ion of melittin in reactions with dimethlyamine is used here for illustration. Melittin contains three lysines and two arginines. It is assumed here that the potential sites of protonation are on these basic residues and on the N-terminus. The $(M + 4H)^{4+}$ ion reacts quantitatively, and with a single rate constant, to give the $(M + 3H)^{3+}$ ion and protonated dimethylamine under ion trap conditions. From these observations it can be concluded that no set of protonated sites gives a coulombic field small enough to make the $\Delta\varepsilon$ value for proton transfer unfavorable. That is, no combination of protonation sites within the quadruply charged ion is stable with respect to proton transfer with dimethylamine. The fact that there appears to be a single rate constant suggests, but does not prove, that a single, most stable structure of $(M + 4H)^{4+}$ makes up the initial parent ion population. This scenario is reasonable given the fact that the kinetic experiment does not begin until well over 100 ms have elapsed from the time the molecule leaves the electrospray needle to the time the $(M + 4H)^{4+}$ ion is isolated for study. Thus, there is time for intramolecular proton transfer to occur.

The $(M + 3H)^{3+}$ ion does not react with dimethylamine either by proton transfer or by clustering, which would seem to indicate that no $(M + 3H)^{3+}$ structure has protonated sites in close enough proximity to make proton transfer a favored reaction. In studying the ion/molecule reactions of the CID product ions from the $(M + 4H)^{4+}$ parent, it was found that no doubly or singly charged product ions reacted with dimethylamine. These observations are significant as they indicate that dimethylamine is probably not sufficiently basic to remove a proton from any of the likely protonation sites and that all of the doubly charged ions observed can maintain a separation of the charges sufficient to preclude proton transfer.

Interestingly, of the two most abundant triply charged ions observed from CID of the $(M + 4H)^{4+}$ parent, the Y_{24}^{3+} and Y_{13}^{3+} products, only the Y_{13}^{3+} ion was observed to react. Of the three triply charged ions examined in this study, the Y_{24}^{3+} and $(M + 3H)^{3+}$ ions did not react with dimethylamine. From all of the observations listed here, assuming that only lysines, arginines, and N-termini are protonated, it is apparent that the only ions that react are those in which two protons are present on the adjacent lys-arg-lys-arg residues (labeled sites C through F in Figure 28). The $(M + 3H)^{3+}$ and Y_{24}^{3+} ions can have a proton on their respective N-termini, on the lysine at site B of Figure 28, and one somewhere in the sequence of residues labeled C to F. With this arrangement, therefore, it is reasonable to conclude that the $(M + 4H)^{4+}$ species is protonated on the N-terminus, on the lysine at site B, and twice in the C to F sequence, most likely on the lysine at site C and the arginine at site F. This conclusion is hardly definitive without some independent confirmation. However, it is clear that information obtained from ion/molecule reactions permits the elimination of many of the possible combinations of protonated sites, and allows a most likely structure to be postulated for further experiments.

Another possible application for ion/molecule reactions is the determination of product-ion charge state in MS/MS spectra.[49] As demonstrated above in the section describing high resolution MS/MS analysis of peptides and proteins, product-ion charge state can be determined from the spacings of the isotope peaks which can require resolving powers of as low as a few thousand to as high as many tens of thousands, depending upon the mass of the product ion of interest. As indicated above, high resolution experiments require relatively long scan times and are, therefore, most appropriate when small regions of the spectrum are of interest. Ion/molecule reactions provide a possible alternative to high mass resolution measurements.

Product-ion charge state determination by an ion/molecule reaction is based upon changing the mass or charge or m/z of the product ion in a known fashion. In the case of proton transfer, the mass of the product

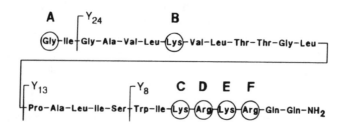

FIGURE 28
The sequence for melittin with the potential sites of protonation indicated and the locations of cleavages that yield various Y-type ions of interest.

ion upon proton transfer decreases by 1 Da and its charge decreases by 1. With two m/z measurements, one before reaction and one after reaction, product-ion charge can be determined in the same way as ion charge is determined from an electrospray mass spectrum.[50] When the product ion forms a proton-bound dimer with the base, the mass of the ion changes by the mass of the base while the charge remains constant. The charge of the product ion, n, is determined by

$$n = m_B / \left(\left| (m/z_1) - (m/z_2) \right| \right) \tag{10}$$

where m_B is the mass of the base and (m/z_1) and (m/z_2) are the m/z ratios of the CID product ion and the adduct ion resulting from clustering, respectively.

Figure 29 shows the result of an MS^3 experiment involving the Y_{13}^{3+} ions from the $(M + 4H)^{4+}$ parent ion of melittin discussed above. Figure 29a shows the result after CID of the parent ion and isolation of the Y_{13}^{3+} product ion followed by a reaction period of 20 ms in the presence of about 2×10^{-7} torr of 1,6-diaminohexane. Figure 29b shows the results from the same experimental sequence, but with 150 ms allowed for re-

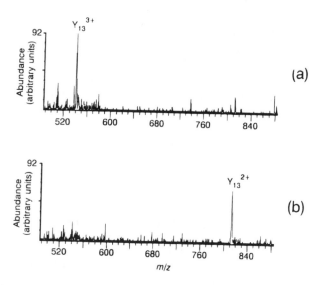

FIGURE 29

Results of an MS/MS/MS experiment following the sequence $(M + 4H)^{4+} \rightarrow Y_{13}^{3+} \rightarrow ?$ where M = melittin. Collisional activation of the $(M + 4H)^{4+}$ ion yields the Y_{13}^{3+} ion, which is then subjected to reaction with dimethylamine. The upper spectrum (a) results when <20 ms are available for reaction, whereas the lower spectrum (b) results after 100 ms are available for reaction. (Reprinted with permission from Ref. 49. Copyright 1991, American Chemical Society.)

action. An essentially quantitative shift to the Y_{13}^{2+} ion is observed, clearly indicating the charge state of the product ion. A mass resolution of about 2000 would suffice to obtain charge state information via the isotope peak spacings, in this case. However, the determination of the charge state for a product ion from the $(M + 20H)^{20+}$ parent ion from horse skeletal muscle myoglobin has also been done via a proton transfer reaction would require a mass resolution of almost 20,000.[49]

An advantage of the use of proton transfer or clustering for charge state determination is that much lower mass resolving power is required to measure the m/z shift between reactant and product ions than for measuring m/z differences in the isotope peaks. The experiment requires a reaction time on the order of 50 to 100 ms, thereby increasing the overall time of analysis. If all of the product ions were to be subjected to a MS^3 study in turn, the time of analysis might rival that required for high resolution measurement of all of the product ions. However, all of the ions that react with the base will do so in one reaction time period. Therefore, much of the information may be obtainable from a comparison of the product ion spectrum acquired in the absence of base (or at a very short reaction time) and one acquired following a single reaction period. In any case, a weakness of this approach is that only some ions react with the base. To date, we have found no gaseous base sufficiently strong to deprotonate all multiply charged protein and protein fragment ions. As discussed at some length above, the ions most likely to react are those with relatively high coulombic fields. Such a situation is most likely to prevail with the most highly charged ions. The most highly charged species tend to be larger molecules or fragments. Ion/molecule reactions are most likely to be of use in those cases requiring the highest mass resolving powers to resolve the isotope peaks. Therefore, the two approaches may prove to be complementary. Ion/molecule reactions may be attempted first to acquire as much charge state information as can be revealed followed by high resolution measurements on product ions that do not react.

B. Mass Spectrometry and MS/MS of Oligonucleotides

Peptides and proteins are by far the subject of most electrospray studies involving biopolymers. It is for this reason in part that such heavy emphasis was given to these species in this chapter. Of course, other important classes of biopolymers might be studied by ES/ITMS. Oligonucleotides are among these and a brief account of ion trap work involving relatively small oligonucleotides is given here. In contrast to peptides and proteins, which usually contain more basic residues than acidic residues, the highly acidic phosphodiester linkages make oligonucleotides more amenable to analysis in the negative ion mode than in the positive ion mode. Negative ion studies pose no significant difficulties for ion trap mass analysis provided a negative ion detection system (for example, conversion dynode/electron multiplier combination) is available. Therefore, all of the ion manipulation and mass analysis capabilities of the ion trap illustrated in the preceding sections can, in principle, be applied to multiply charged anions derived from oligonucleotides. However, oligonucleotides are, as a rule, much less robust than peptides and proteins so that some differences in the behavior of ions derived from these classes may be expected under various ion trap conditions. Too little work has been done thus far with oligonucleotides, however, to draw general conclusions on the nature and extent of these differences.

Figure 30 shows the electrospray ion trap mass spectrum acquired at the normal scan speed for the oligomer polydeoxyadenoic acid, $pd[A]_{10}$. Like proteins, several charge states are observed, including, in this case, anions with five, four, three, and two charges, respectively. Unlike proteins, oligonucleotides show extensive incorporation of sodium ions[2,4,50] unless solutions are prepared with care so as to minimize their presence.[51] In this case, there are nine phosphodiester linkages and a phosphate group attached to the 5'-terminus, giving a maximum charge of ten, assuming all acidic protons are removed. Sodium ions can replace protons resulting in clusters of peaks at each charge state comprised of ions with different combinations of protons and sodium ions. The inset to Figure 30 shows an expanded scale for the 4– charge state clearly showing the incorporation of up to seven sodium ions, the maximum number expected. (In the neutral molecule two protons are associated with the 5'-phosphate which when combined with the nine protons associated with the phos-

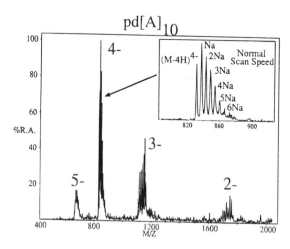

FIGURE 30

ES mass spectrum of $pd[A]_{10}$ acquired using the normal scan speed of the ion trap.

phodiester linkages, lead to 11 possible exchange sites.) Of course, the number of possible sodium exchanges is related inversely to the charge state so that the multiplicity of peaks at a given charge state tends to be largest for the lowest charge states.

Figure 31 shows electrospray mass spectra of pd[A]$_{10}$ over the region of the 4– anions and obtained with scan speeds reduced by a factor of 10 (upper) and by a factor of 100 (lower). The latter spectrum showing a mass resolution of roughly 3300. These results suggest that high resolution mass measurements can be made with multiply charged oligonucleotide anions just as they have been demonstrated with multiply protonated proteins. However, these results are demonstrated with species with a relatively low degree of charging. More highly charged species are expected to be less stable due to increased internal coulombic repulsion. Significant internal excitation of ions is expected under extremely slow scanning resonance ejection conditions which can lead to CID of fragile ions. As dissociation has been observed under such conditions for multiply charged peptides and fragile singly charged species, it could be problematic for either highly charged proteins or highly charged oligonucleotides. However, it is expected to be more likely with the latter.

Relatively little work on the collisional activation of multiply charged oligonucleotide anions has been reported, in contrast to the case for peptides and proteins. Nevertheless, the highly successful MS/MS approaches

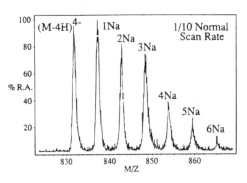

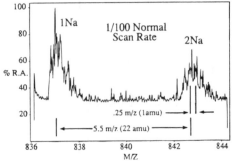

FIGURE 31
ES mass spectra of pd[A]$_{10}$ over narrow portions of the mass range including ions from the 4– charge state. The upper spectrum shows a scan acquired at $\frac{1}{10}$ the normal scan speed and the lower spectrum shows results obtained at $\frac{1}{100}$ the normal scan speed.

to sequencing peptides that have been developed in recent years prompts the consideration that similar capabilities might be possible for oligonucleotides. We reported recently on an ion trap collisional activation study of several relatively small multiply charged deoxy oligomers ($n = 4$ to 8) which indicated that structurally informative fragmentation could be derived from these ions.[52] Some of these results are described here. Since then, we have acquired much more data that show that these species fragment in a highly predictable and informative fashion. The prognosis for sequencing relatively small oligomers ($n < 20$) is very positive. These more recent results are not presented here because oligonucleotide sequencing is not the focus of this chapter, but some of the general conclusions are summarized.

Multiply charged oligonucleotide anions tend to fragment very readily in the ion trap, even under mild collisional activation conditions. This tendency increases with charge state, as expected. Therefore, MS/MS efficiencies tend to be quite high. Furthermore, under relatively gentle collisional activation conditions, both product ions from a fragmentation reaction involving charge separation are generally observed. Under more violent conditions, sequential fragmentations are observed, making identification of complementary ions difficult *a priori*. However, with a knowledge of the strong tendencies in fragmentation that these species show, it is generally a straightforward matter to determine the origins of all of the major ions in the spectrum. Some of these tendencies are illustrated in Figure 32, which shows the MS/MS spectrum acquired from the triply charged anion of $d[A]_4$ under mild collisional activation conditions, and the MS^3 spectrum following the experimental sequence $d[A]_4^{3-} \rightarrow (d[A]_4-A^-)^{2-} \rightarrow ?$.

The MS/MS spectrum illustrates the general observation that the lowest energy fragmentation pathway for a "highly charged" oligonucleotide that contains at least one adenine in the sequence is the loss of charged adenine. (The term "highly charged" is meant to imply that a high percentage of the total maximum charge, assumed to be the sum of the phosphate and phosphodiester linkages in the oligomer, is carried by the ion.) In the case illustrated in Figure 32, of course, all of the bases are adenines, but whenever there are others, loss of charged adenine dominates. The only exceptions to this rule noted thus far occur when only one adenine is found in the sequence and is located at the 3'-terminus, and when the oligomer is phosphorylated at the 5'-end. In the latter case, loss of PO_3^- dominates. When the ion is not highly charged, loss of the base as a neutral is the low energy decomposition. For example, the doubly charged $d[A]_4$ anion loses predominantly neutral adenine as a first step. Thus far, no strong and consistent tendency for the order in which neutral bases are lost in a mixed oligomer has become apparent.

Base loss from the various sugars along the oligomer are competitive processes for the first step in dissociation. However, once one base is lost

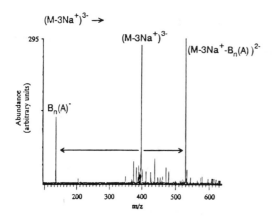

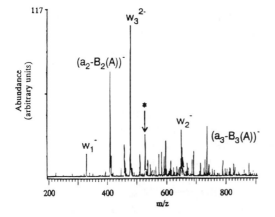

FIGURE 32
MS/MS and MS/MS/MS spectra of $d[A]_4^{3-}$. The MS/MS/MS spectrum follows the sequence $d[A]_4^{3-} \rightarrow (d[A]_4^{3-} -A^-)^{2-} \rightarrow ?$. (Reprinted with permission from Ref. 52. Copyright 1992, Elsevier Science Publishing.)

the next step is not the loss of another base. Rather, the oligomer breaks at the 3′ C-0 bond of the sugar from which the base was lost. This mode of dissociation results in complementary w- and (a–B)-type ions, according to a nomenclature described in the original paper.[52] This cleavage indicates the position of the base in the chain. The MS³ experiment leading to the lower spectrum of Figure 32 illustrates the tendency by collisionally activating the ions resulting from the loss of charged adenine. The major second-generation product ions in this spectrum are due to cleavages at the 3′ C-0 bonds of the three sugars beginning from the 5′-terminus. The major product ion, w_3^{2-}, arises from the loss of the 5′-sugar, whereas the other product ions arise from charge separation reactions yielding the complementary w_2^- and $a_2\text{-}B_2(A)^-$ ions and the complementary w_1^- and $a_3\text{-}B_3(A)^-$ ions from the second and third sugars from the 5′-terminus, respectively. Interestingly, no evidence is apparent for loss of the 3′-adenine. The analogous cleavage for this base loss would result in water loss. However, water loss has been observed to be a significant reaction only for nucleotides. Either the 3'-base is not lost or water loss is significantly less favored than the analogous cleavages of the other 3' C-0 bonds. The energetics are expected to differ for loss of water, but we have also found evidence that the 3'-base shows a lesser tendency for cleavage,[53] at least for highly charged ions. In any case, the cleavages in Figure 32 are sufficient to sequence the oligomer in this trivial case.

We have found that the cleavage at the 3' C-0 bond of the sugar from which the base is lost is independent of the identity of the base, and that oligomer fragments that do not contain adenine show effective competition for loss of many if not all of the remaining bases. Therefore, complete sequencing of oligonucleotides, at least up to $n = 12$, appears to be feasible via a set of well-chosen MS/MS and MS^n experiments.[54] The ion trap appears to be particularly well suited to performing these types of experiments due both to its tandem-in-time capabilities and to the nature of ion trap collisional activation. Multiply charged oligonucleotides tend to undergo extensive sequential decomposition under relatively violent collisional activation conditions. This can make spectral interpretation difficult. However, the ion trap affords unusually good control over the approach to the first decomposition threshold of an ion. This capability is valuable in controlling the extent to which these species fragment, thereby enhancing the value of the ion trap for sequencing oligonucleotides.

V. CONCLUSIONS AND PROGNOSIS

Only a few years ago it would have been almost unimaginable that structural information from proteins or oligonucleotides could be obtained with a quadrupole ion trap. However, the advent of electrospray and the remarkably rapid evolution of the ion trap as a high performance analytical mass spectrometer have made these and many other capabilities possible. The electrospray/ion trap combination with its high m/z range, MS^n capabilities, capability for high resolution, high sensitivity, and small relative size and cost, will be an important analytical tool. The potential for biological research is particularly apparent from this chapter's emphasis on multiply charged biopolymers. However, many other areas, such as environmental, clinical, and forensic analysis, will benefit from this combination. For many applications the ES/ITMS has already moved past the curiosity stage and can be considered seriously as the method of choice. That is not to say, however, that all challenges have been met. For example, obtaining high mass accuracy on a routine basis must still be addressed, and other limitations imposed by ion-ion interactions merit attention. Furthermore, improvements in sensitivity, resolution, speed, and other figures of merit can, and almost certainly will, be made. There are many opportunities to improve the existing state of the art. Nevertheless, the electrospray/ion trap results discussed here and elsewhere suggest clearly that this combination will enjoy widespread employment both in research and in routine analysis.

ACKNOWLEDGMENTS

All of the results reported here from Oak Ridge National Laboratory were obtained with support from the U.S. Department of Energy, Office of Basic Energy Sciences under Contract DE-AC05-84OR21400 with Martin Marietta Energy Systems, Inc. J. C. Schwartz acknowledges the collaborative efforts of C. G. Edmonds of Battelle Pacific Northwest Laboratories and his colleagues in the Finnigan Research and Engineering Group.

REFERENCES

1. Fenn, J.B.; Mann, M.; Meng, C.K.; Whitehouse, C.M. *Science.* 1990, 246, 64.
2. Smith, R.D.; Loo, J.A.; Edmonds, C.G.; Barinaga, C.J.; Udseth, H.R. *Anal. Chem.* 1990, 62, 882.
3. Fenn, J.B.; Mann, M.; Meng, C.K.; Wong, S.F.; Whitehouse, C.M. *Mass Spectrom. Rev.* 1990, 9, 37.
4. Smith, R.D.; Loo, J.A.; Ogorzalek Loo, R.R.; Busman, M.; Udseth, H.R. *Mass Spectrom. Rev.* 1991, 10, 359.
5. Mann, M. *Org. Mass Spectrom.* 1990, 25, 575.
6. Huang, E.C.; Wachs, T.; Conboy, J.J.; Henion, J.D. *Anal. Chem.* 1990, 62, 713A.
7. Van Berkel, G.J.; McLuckey, S.A.; Glish, G.L. *Anal. Chem.* 1991, 63, 2064.
8. Van Berkel, G.J.; McLuckey, S.A.; Glish, G.L. *Anal. Chem.* 1992, 64, 1586.
9. Wong, S.F.; Meng, C.K.; Fenn, J.B. *J. Phys. Chem.* 1988, 92, 546.
10. Meng, C.K.; Mann, M.; Fenn, J.B. *Z. Phys. D.* 1988, 10, 361.
11. Louris, J.N.; Cooks, R.G.; Syka, J.E.P.; Kelley, P.E.; Stafford, G.C., Jr.; Todd, J.F.J. *Anal. Chem.* 1987, 59, 1677.
12. Louris, J.N.; Brodbelt-Lustig, J.S.; Cooks, R.G.; Glish, G.L.; Van Berkel, G.J.; McLuckey, S.A. *Int. J. Mass Spectrom. Ion Processes.* 1990, 96, 117.
13. McLuckey, S.A.; Glish, G.L.; Van Berkel, G.J. *Int. J. Mass Spectrom. Ion Processes.* 1991, 106, 213.
14. Kaiser, R.E., Jr.; Cooks, R.G.; Moss, J.; Hemberger, P.H. *Rapid Commun. Mass Spectrom.* 1989, 3, 50.
15. Kaiser, R.E., Jr.; Louris, J.N.; Amy, J.W.; Cooks, R.G. *Rapid Commun. Mass Spectrom.* 1990, 3, 225.
16. Schwartz, J.C.; Syka, J.E.P.; Jardine, I. *J. Am. Soc. Mass Spectrom.* 1991, 2, 198.
17. Williams, J.D.; Cox, K.A.; Cooks, R.G.; Kaiser, R.E., Jr.; Schwartz, J.C. *Rapid Commun. Mass Spectrom.* 1991, 5, 327.
18. Mordehai, A.; Henion, J.D. *Proc. 40th ASMS Conf. Mass Spectrom. Allied Topics.* Washington, D.C., 1992; p. 197.
19. Van Berkel, G.J.; Glish, G.L.; McLuckey, S.A. *Anal. Chem.* 1990, 62, 1284.

20. Hail, M.E.; Schwartz, J.C.; Mylchreest, I.C.; Seta, K.; Lewis, S.; Zhou, J.; Jardine, I.; Liu, J.; Novotny, M. *Proc. 38th ASMS Conf. Mass Spectrom. Allied Topics.* Tucson, AZ, 1990; p. 353.
21. Jonscher, K.R.; Currie, G.J.; McCormack, A.L.; Yates, J.R., III. *Proc. 40th ASMS Conf. Mass Spectrom. Allied Topics.* Washington, D.C., 1992; p. 701.
22. Voyksner, R.D.; Lin, H.Y.; Pack, T. *Proc. 39th ASMS Conf. Mass Spectrom. Allied Topics.* Nashville, TN, 1991; p. 1643.
23. McLuckey, S.A.; Glish, G.L.; Asano, K.G. *Anal. Chim. Acta.* 1989, 225, 25.
24. Asano, K.G.; McLuckey, S.A.; Glish, G.L. *Spectrosc. Int. J.* 1990, 8, 191.
25. McLuckey, S.A.; Glish, G.L.; Asano, K.G.; Grant, B.C. *Anal. Chem.* 1988, 60, 2220.
26. Smith, R.D.; Loo, J.A.; Barinaga, C.J.; Udseth, H.R. *J. Am. Soc. Mass Spectrom.* 1990, 1, 53.
27. Schoen, A.E.; Syka, J.E.P. *Proc. 36th ASMS Conf. Mass Spectrom. Allied Topics.* San Francisco, 1988; p. 843.
28. McLuckey, S.A.; Goeringer, D.E.; Glish, G.L. *J. Am. Soc. Mass Spectrom.* 1991, 2, 11.
29. Ikonomou, M.G.; Blades, A.T.; Kebarle, P. *Anal. Chem.* 1990, 62, 957.
30. Jayaweera, P.; Blades, A.T.; Ikonomou, M.G.; Kebarle, P. *J. Am. Chem. Soc.* 1990, 112, 2452.
31. Yamashita, M.; Fenn, J.B. *J. Phys. Chem.* 1984, 88, 4451.
32. Bruins, A.P.; Covey, T.R.; Henion, J.D. *Anal. Chem.* 1987, 59, 2642.
33. Katta, V.; Chowdhury, S.K.; Chait, B.T. *J. Am. Chem. Soc.* 1990, 112, 5348.
34. Van Berkel, G.J.; McLuckey, S.A.; Glish, G.L. *Anal. Chem.* 1991, 63, 1098.
35. Johnson, J.V.; Britton, E.D.; Yost, R.A.; Quirke, J.M.E.; Cuesta, L.L. *Anal. Chem.* 1986, 58, 1325.
36. Loo, J.A.; Udseth, H.R.; Smith, R.D. *Rapid Commun. Mass Spectrom.* 1989, 2, 207.
37. McLuckey, S.A.; Goeringer, D.E.; Glish, G.L. *Anal. Chem.* 1992, 64, 1455.
38. Van Berkel, G.J.; Ramsey, R.S.; McLuckey, S.A.; Glish, G.L. *Proc. 40th ASMS Conf. Mass Spectrom. Allied Topics.* Washington, D.C.; 1992; p. 711.
39. Roepstorff, P.; Fohlman, J. *Biomed. Mass Spectrom.* 1984, 11, 601.
40. Schwartz, J.C.; Jardine, I. *Rapid Commun. Mass Spectrom.* 1992, 6, 313.
41. McLuckey, S.A.; Van Berkel, G.J.; Glish, G.L. *J. Am. Chem. Soc.* 1990, 112, 5668.
42. Ogorzalek Loo, R.R.; Loo, J.A.; Udseth, H.R.; Smith, R.D. *Rapid Commun. Mass Spectrom.* 1992, 6, 159.
43. Winger, B.E.; Light-Wahl, K.J.; Smith, R.D. *J. Am. Soc. Mass Spectrom.* 1992, 3, 624.
44. Ikonomou, M.G.; Kebarle, P. *Int. J. Mass Spectrom. Ion Processes.* 1992, 117, 283.
45. Feng, R.; Konishi, Y. *Proc. 40th ASMS Conf. Mass Spectrom. Allied Topics.* Washington, D.C., 1992; p. 1635.
46. Suckau, D.; Shi, Y.; Quinn, J.P.; Senko, M.W.; Zhang, M.-Y.; McLafferty, F.W. *Proc. 40th ASMS Conf. Mass Spectrom. Allied Topics.* Washington, D.C., 1992; p. 477.
47. McLuckey, S.A.; Asano, K.G.; Glish, G.L.; Bartmess, J.E. *Int. J. Mass Spectrom. Ion Processes.* 1991, 109, 171.
48. Rockwood, A.L.; Busman, M.; Smith, R.D. *Int. J. Mass Spectrom. Ion Processes.* 1991, 111, 103.
49. McLuckey, S.A.; Glish, G.L.; Van Berkel, G.J. *Anal. Chem.* 1991, 63, 1971.
50. Covey, T.R.; Bonner, R.F.; Shushan, B.I.; Henion, J.D. *Rapid Commun. Mass Spectrom.* 1988, 2, 249.
51. Stults, J.T.; Marsters, J.C. *Rapid Commun. Mass Spectrom.* 1991, 5, 359.
52. McLuckey, S.A.; Van Berkel, G.J.; Glish, G.L. *J. Am. Soc. Mass Spectrom.* 1992, 3, 60.
53. Habibi-Goudarzi, S.; McLuckey, S.A. unpublished results, Oak Ridge National Laboratory, 1992.
54. McLuckey, S.A.; Van Berkel, G.J.; Ramsey, R.S.; Glish, G.L. *Proc. 40th ASMS Conf. Mass Spectrom. Allied Topics.* Washington, D.C., 1992; p. 537.

PART 3

Ion Structure Differentiation in an Ion Trap

Chapter 4

EVALUATION OF THE POLARIZABILITY OF GASEOUS IONS

Olga Bortolini and Pietro Traldi

CONTENTS

I.   Introduction 146
     A.   Mass Shifts Observed in the Mass Range
          Extension Mode 147
     B.   Mass Shifts Observed in Normal Ion Trap Mass
          Spectrometer Operation 147
     C.   Mass Shifts Observed for Isobaric Ions 147

II. Influence of Instrumental Parameters on Mass
     Displacements 149
     A.   Space-Charge Effects 150
     B.   Gas Phase Ion/Molecule Interactions 150
     C.   Supplementary Radiofrequency Voltage Effects 150
          1.   Axial Modulation of Molecular Ions of
               Chloronitrobenzene Isomers 151
          2.   Axial Modulation of Daughter Ions of
               Chloronitrobenzene Isomers 151

III. Considerations of Polarizability 153
     A.   Total Polarizability 153
     B.   Frequency Dependence of Polarizability 154

0-8493-4452-2/95/$0.00+$.50

IV.   Correlations between Mass Displacements and
      Polarizability 155
      A.   Mass Displacements of Alkylbenzenes 155
      B.   Mass Displacements of *p*-Alkyl-Substituted
           Pyridines 156
      C.   Mass Displacements of Variously Substituted
           Benzenes 157
      D.   Mass Displacement and the Dipole Moment of the
           Neutral Precursor 158

V.   Conclusion   159

References 159

I. INTRODUCTION

The software developed for the Finnigan MAT Ion Trap Mass Spectrometer (ITMS™) is built so that the mass scale is subdivided by tickmarks spaced by 1 u.[1] The mass value m is assigned to a peak when this is found between the two tickmarks corresponding to m and $(m + 1)$. Thus, for example, when the mass calibration procedure is performed, the peaks arising from the reference compound (FC-43) at m/z 69, 131, 219, 264, 414, 502, and 614 must be found between the tickmarks labeled 69 to 70, 131 to 132, 219 to 220 etc. The peak corresponding to m/z 502 is shown in Figure 1. As is seen in this figure, and as is observed usually, the centroid of the peak corresponding to an ion of mass m Da is very close to the $(m + 0.5)$ mass value.

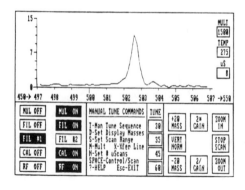

FIGURE 1
Full-scale adjustment display of the m/z 502 peak obtained from the calibration compound, FC-43.

A. Mass Shifts Observed in the Mass Range Extension Mode

In general, this software functions properly and well, leading to reliable and reproducible unit mass assignments. However, Cooks and co-workers[2] reported in 1991 that significant mass shifts were found when using axial modulation for mass range extension. The shifts were found to be dependent on the frequency and on the amplitude of the supplementary voltage applied between the end-cap electrodes. The mass peaks were shifted to lower apparent masses, and the magnitudes of the shifts were observed, in some cases, to be as much as 60 Da.

B. Mass Shifts Observed in Normal Ion Trap Mass Spectrometer Operation

In a further study it was shown that such mass displacements can be detected also, if only to a minor extent, in the usual mass range of an ITMS.[3] Mass displacements from the expected instrumental value (m + 0.5) were observed with a variety of odd- and even-electron cations. The magnitudes and signs of these mass displacements were found to be dependent on the starting scan mass of the scan program, i.e., on the q_z value of the ion under investigation and prior to its ejection.[3] As an example, Figure 2a shows the trend of the mass displacement (δ_m) as a function of the q_z value at which the xenon cation (m/z 129, corresponding to the ^{129}Xe isotope) lies in the stability diagram immediately prior to ejection. At low q_z values, strong mass defects are observed; the magnitude of the mass defect diminishes slowly with increasing q_z value and becomes constant for $q_z \geq 0.4$. Under these conditions, the expected instrumental value of 129.50 Da is found, corresponding to a δ_m value equal to zero. In Figure 2b the trend of δ_m vs. q_z for the nitrobenzene molecular cation (m/z 123) is shown. As in the previous case, negative mass shifts are found for low q_z values; the mass shift becomes positive for $q_z \geq 0.4$ and becomes constant for $q_z \geq 0.5$, whereupon a significant mass displacement equal to 0.18 Da is found.

C. Mass Shifts Observed for Isobaric Ions

Such behavior could be ascribed, in principle, to many instrumental factors as, for example, space-charge effects, but further experiments rule out this hypothesis. In particular, it was found that isobaric ions of the same elemental composition but of different structure and introduced into the ion trap in similar concentrations exhibited different δ_m values (they were detected at different mass values).[4,5] To illustrate this mode of behavior, the δ_m vs. q_z trends for the M$^+$ ions of o- and p-trifluoromethyl

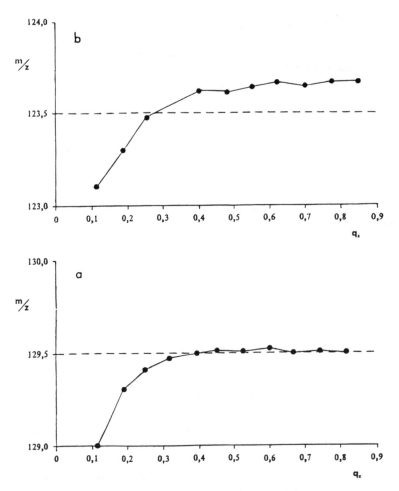

FIGURE 2
Mass/charge ratio as a function of q_z. (a) ^{129}Xe; (b) the molecular cation of nitrobenzene.

benzonitrile are shown in Figure 3. It is notewothy that the δ_m values at which the plateau is reached for each of the two cases, 0.40 Da for the o-isomer and 0.28 Da for the p-isomer, are clearly different.

Mass analysis in the ion trap, under mass-selective ion instability mode conditions, is governed by the equation[6,7]

$$m/z = 4V \left(q_{eject} r_0^2 \Omega^2 \right) \tag{1}$$

where V is the zero-to-peak amplitude of the radiofrequency (RF) voltage applied to the ring electrode, q_{eject} is the value of the q_z Mathieu pa-

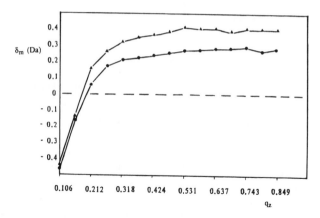

FIGURE 3

Mass displacement, δ_m, of isomeric trifluoromethylbenzonitrile molecular cations as a function of q_z: (▲) *ortho-*; (●) *para-*.

rameter at which the ions are ejected, Ω is the angular component of the RF drive frequency, and r_0 is the internal radius of the ring electrode ($r_0 = 1$ cm).

From the relationship expressed by Equation 1, it follows that the ion trap is an effective mass analyzer whose behavior must be independent of any physico-chemical property of the ions under analysis. This assertion, however, seems to be contradicted by the experimental observations reported above, particularly in the case of isobaric ions. In fact, such behavior must necessarily be ascribed to a physico-chemical property of the ions to which the ITMS™ is responsive. The fundamental importance of this assumption and the operational complexity of the ITMS™ clearly indicated that further experimental work should be undertaken in order to determine if, when, and how the different operational conditions can affect the δ_m measurements.

II. INFLUENCE OF INSTRUMENTAL PARAMETERS ON MASS DISPLACEMENTS

The occurrence of mass shifts could, in principle, originate from many factors either instrumental or related to the physical properties or reactivity of the ionic species under study. Experiments were undertaken in order to investigate four such aspects: these aspects were space-charge effects, gas-phase ion/molecule interactions, supplementary RF voltage effects, and RF and direct current (DC) field effects.

A. Space-Charge Effects

The perturbations due to space-charge effects during ion confinement may cause shifts of the peak position.[6,8] To establish whether the mass displacements were affected by this phenomenon, several experiments were carried out which consisted of reducing systematically the number of ions with respect to the ionic species being investigated. For this purpose, the mass displacements of a model ion species were measured after its isolation from all other ion species in the trap, and the results were compared to the δ_m value found by simple mass analysis.[5] The ionic species of m/z 152, originating from p-trifluoromethylbenzonitrile by loss of F· was selected as a model, and its mass displacements were measured after isolation by both the apex method[9] and the two-step isolation procedure.[10] The two results, $\delta_m = 0.12$ and 0.13 ± 0.02 Da, respectively, should be compared with the mass displacement found in the absence of ion isolation, on the plateau of the δ_m vs. q_z curve. For this ion species, the plateau value was found to be $\delta_m = 0.12$ Da.

These results proved that space-charge effects do not have an appreciable influence on the δ_m measurements. Furthermore, the selected ions experience operating conditions during ion isolation that are somewhat different than those employed during simple mass analysis; in particular, a stronger RF field associated with a DC component is used in the ion isolation procedure. However, the measured value for δ_m remained practically unchanged, proving that the history of the ion before ejection had no influence on mass displacement measurements.

B. Gas Phase Ion/Molecule Interactions

Chemical interactions of some kind may be responsible for this behavior. The effect of possible chemical interactions among the ions stored within the ion trap was investigated using the [M-H]+ ion of trifluoromethylbenzene.[5] A reaction time of length varying from 100 to 2×10^5 μs, was introduced before the main RF ramp. Chemical interactions should be time dependent, and a change in this parameter should be reflected by a variation in δ_m. As shown in Figure 4, negligible differences in the δ_m values were found for increasing interaction times, thus ruling out possible chemical interactions as responsible for the observed mass shifts.

C. Supplementary Radiofrequency Voltage Effects

Alternatively, the phenomenon of mass displacements could be rationalized in terms of some physico-chemical property of the ions under study and the RF field(s) present in the ion trap. The consequence of this

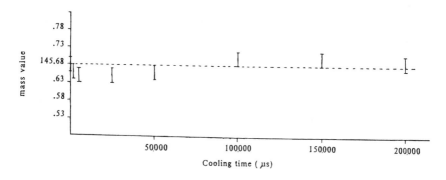

FIGURE 4

Variation of the mass value as a function of cooling (or reaction) time for the [M-H]⁺ ion of trifluoromethylbenzene.

interaction leads to a small but detectable power consumption of the RF fields, which is reflected in the ejection of the ions at different q_z values, i.e., at different values of the RF drive voltage, V_{RF}.

1. Axial Modulation of Molecular Ions of Chloronitrobenzene Isomers

In order to verify whether such interaction also could be present with the supplementary RF voltage (tickle voltage), a number of experiments were carried out using axial modulation. The o-, m-, and p-chloronitrobenzene isomers were selected as probes for this experiment, and the corresponding molecular cations were isolated using the two-step procedure and irradiated by an auxiliary RF voltage of increasing amplitude. It was found that while the initial signal intensity of the M⁺ ion of the para isomer was reduced to 10 percent of the initial intensity using a tickle voltage amplitude of 1800 mV, in order to reduce the intensities of the m- and o-isomers by a similar amount, the required tickle amplitudes were 1900 and 2200 mV, respectively, as shown in Figure 5.

2. Axial Modulation of Daughter Ions of Chloronitrobenzene Isomers

The interaction of an isolated ionic species with the auxiliary RF field was confirmed further by a comparison of the mass displacements found for the same ion species generated both by electron impact and by collision, as described below. The molecular ions of o-, m-, and p-chloronitrobenzene were isolated by the two-step procedure and irradiated at the fundamental axial secular frequency with a supplementary RF voltage so as to produce daughter ions. The mass displacements of the [M-NO]⁺ ionic species obtained by resonance excitation were measured at different q_z values for all the isomers, and the results were compared with the

δ_m values found in the usual mass analysis, i.e., in the absence of irradiation. The results of these experiments are reported in full in Table 1. An increase in δ_m as the magnitude of q_z was increased was observed in all cases; however, for a given isomer and at a given value of q_z, the mass displacement found for the collisionally generated ion was consistently greater than that measured in the absence of irradiation. The results prove that a stronger interaction must occur in those experiments in which the ions were generated collisionally, and that this interaction occurs with the supplementary RF voltage.

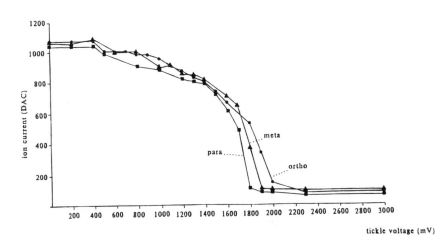

FIGURE 5
Signal intensities (in DAC units) of the isomeric molecular cations of chloronitrobenzene as a function of supplementary RF (or tickle) voltage amplitude: (●) *ortho*-; (▲) *meta*-; (■) *para*-

TABLE 1

Mass Displacement, δ_m, Measurements at Several q_z Values for the Isomeric [M-NO]⁺ Ions Generated by Electron Impact of *o*-, *m*- and *p*-Chloronitrobenzene and in Collisional Experiments with the Molecular Cations Using Tickle Voltage Activation.

Isomer	Mass displacement (Da)				
	Electron impact, q_z		Collisional generation, q_z		
	0.03	0.80	0.50	0.66	0.80
Ortho	0.05[a]	0.07	0.12	0.14	0.16
Meta	0.05	0.05	0.10	0.13	0.14
Para–	0.10	0.03	0.00	0.12	0.06

[a] The uncertainty in all values of Δ_m is ±0.02.

It is particularly noteworthy that the magnitudes of the mass displacements of the isomeric disubstituted benzene compounds examined were always in the order *ortho > meta > para*.

Upon consideration of the results presented above, the physico-chemical property interacting with the RF field(s) was recognized to be the polarizability because, by a process of elimination, it is the only property that can account for different mass displacements for isobaric ions of identical elemental composition but of different structure. This conclusion is consistent with the experimental observation that the magnitude of δ_m increases as the dipole moment of the analyte increases. Additional support for this hypothesis came also from the behavior observed for similar isobaric ions (of identical elemental composition but of different structures) in a quadrupole mass filter.[11] Clear differences were found in the mass shifts for different isomers, proving that the mass displacement effect observed with the ion trap is but a general phenomenon that occurs when ions experience a quadrupole field.

III. CONSIDERATIONS OF POLARIZABILITY

The polarizability, α_0, of an atom or molecule describes the response of the analyte to the action of an external field. In a dielectric medium, the polarizability is related to the macroscopic or molar polarization, P, by the expression[12,13]

$$\alpha_0 = 3P/(4\pi N_0)$$

(2)

where N_0 is Avogadro's number.[14]

A. Total Polarizability

Depending on the physical characteristics of the molecules, the polarizability and, hence, the polarization can be expressed as the sum of three contributions: electronic, orientational (dipolar), and atomic. The electronic polarization, P_e, is due to the deformation of the outer electron shell of the molecule; the atomic polarization, P_a, refers to the displacement of the positive nuclei of the atoms composing the molecule or, more generally, to the displacement of atoms or groups of atoms within the molecule; the orientational component, P_o, arises when the analyte possesses a permanent electric dipole moment which tends to align along the field direction. An electric field always induces a dipole moment in a molecule, whether or not the molecule has a permanent dipole moment. Then the total polarization may be expressed as

$$P = P_e + P_a + P_o$$

(3)

B. Frequency Dependence of Polarizability

In the optical frequency range, the contribution to P arises almost entirely from the electronic polarization. The orientational and atomic contributions are small at high frequencies because of the inertia of the molecules and ions.

At lower frequencies (infrared, for example), the oscillations of the field are sufficiently slow for the atoms to follow; consequently, the polarization is the sum of the electronic and atomic contributions. In the relatively slow oscillations of an alternating electric field in the range of 10^3 to 10^6 Hz, the permanent dipoles have sufficient time to adopt the statistical distribution assumed by Debye, and the total polarization is given by Equation 3. The frequency dependence of the different contributions to the polarization is shown in Figure 6.

Experiments carried out on the variation with temperature of the dielectric constant of a gas indicate that the atomic contribution is small, even for quite large molecules. Therefore, P_a can be neglected in comparison with P_o or, alternatively, P_a may be assumed to be a constant fraction of the electronic polarization P_e. The value of P_e can be calculated from the refractive index, n, extrapolated to infinite wavelength ($\lambda \to \infty$)

$$P_e = [(n^2 - 1) M]/[(n^2 + 2) \rho] \tag{4}$$

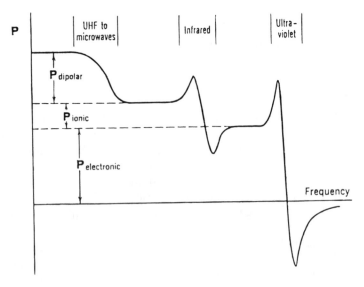

FIGURE 6
Frequency dependence of the dipolar, ionic, and electronic components of the total polarization, P.

where M is the molecular weight and ρ is the density. According to Debye,[13] however, the value of P_e calculated from the refractive index, n_D, of the sodium line gives the correct value for P_e, on the assumption that $P_a = 0$. In the low frequency region (VHF, UHF, microwaves) the total polarization is

$$P = P_e + P_o \tag{5}$$

and P_e is calculated from the expression

$$P_e = \left[\left(n_D^2 - 1 \right) M \right] / \left[\left(n_D^2 + 2 \right) \rho \right] \tag{6}$$

The values obtained using Equation 6 are greater than those obtained by extrapolation to zero frequency, and this difference compensates partially for the neglect of the atomic polarization.

IV. CORRELATIONS BETWEEN MASS DISPLACEMENTS AND POLARIZABILITY

For commercially available ITMS™ instruments, the frequency utilized for the RF drive potential is 1.1 MHz. Therefore, in this frequency region, the contributions to the total polarizability (or polarization) are due to the electronic component and, in the presence of permanent dipoles, also to the orientational component.

A. Mass Displacements of Alkylbenzenes

From a physical and qualitative point of view, the induction of the electronic polarizability and the alignment of permanent dipoles along the field direction should be the same for both neutral and ionic analytes. On this assumption, the phenomenon of mass displacement was investigated[15] using several aromatic monosubstituted molecular cations $[C_6H_5\text{-}Y]^+$. The analytes were divided into two sets depending on the nature of the Y-substituent. The first group of ions was composed of alkylbenzenes of increasing linear chain length, i.e., $[C_6H_6]^+$, $[C_6H_5\text{-}CH_3]^+$, $[C_6H_5\text{-}CH_2CH_3]^+$, $[C_6H_5\text{-}(CH_2)_2CH_3]^+$, $[C_6H_5\text{-}(CH_2)_5CH_3]^+$, each possessing a small and similar permanent dipolar polarizability. The contribution to α_0 is due mainly to the electronic polarizability. Accordingly, a direct linear relationship between the experimental value for δ_m and the value for α_0 calculated by Equations 2 and 6 was found, where the dipolar component represents a small, constant fraction affecting only the intercept. The linear trend of δ_m vs. α_0 is shown in Figure 7. The trend shown here also can accommodate other substrates such as styrene, isopropylbenzene, biphenyl, and o- and p-xylene ions.

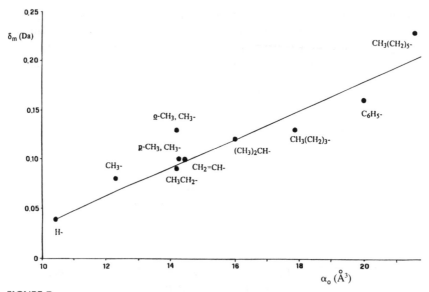

FIGURE 7

Mass displacements, δ_m, for $[C_6H_5-Y]^{+\cdot}$ ions as a function of the electronic polarizability, α_0, calculated from Equations 2 and 6. The Y-substituents are shown.

The observed variation of the mass displacement of the molecular ion with the polarizability of the neutral precursor constitutes strong support for the following hypothesis: the origin of the mass displacement is due to an interaction of the dipole moment induced in the ion by the RF field and the RF field itself. The linearity of this variation is supportive of the proposal that the classical treatment described above, of a dielectric medium composed of atoms or neutral molecules plunged into an electric field, can be extended to gaseous ions without the necessity to make significant conceptual changes.

The forces affecting the electron distribution in a neutral molecule placed in an electric field should, in fact, be similar to those acting on a charged species, at least for those ions that are able to distribute the vacancy of an electron by delocalization effects.[15]

B.   Mass Displacements of *p*-Alkyl-Substituted Pyridines

Further support for this hypothesis is found in results obtained with a second group of compounds composed of *p*-alkyl-substituted pyridines. Once again, a linear dependence of the experimental mass shift, δ_m, upon the magnitude of the polarizability of the neutral precursor, α_0, was found,[16] similar to that observed for the molecular ions of alkylbenzenes. For the

p-alkyl-substituted pyridines, the dipolar polarizability appears to be a constant and rather small fraction of the total polarizability; this property is manifested in the intercept only.

C. Mass Displacements of Variously Substituted Benzenes

The measured mass displacements for a collectivity of variously substituted benzenes, each possessing a permanent dipolar polarizability, plotted against the electronic polarizability calculated according to Equations 2 and 6 are shown in Figure 8. The sloping line included in this plot represents the trend observed for alkylbenzene molecular cations, and shown previously in Figure 7. The presence of the dipolar component of the polarizability in the variously substituted benzenes is indicated clearly here; the sloping line shows the dependence of the mass shift upon electron polarizability, and the additional dipolar component of this second set of molecular cations causes all of the mass displacements to lie above the sloping line. In addition, analytes with a high permanent dipolar component show large departures from the linear trend depicted by the sloping line.

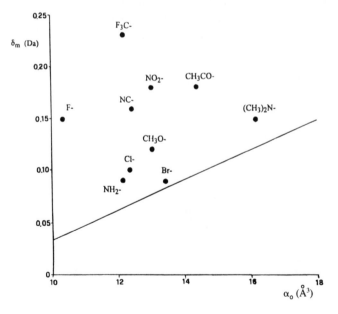

FIGURE 8

Mass displacements, δ_m, for a set of substituted benzene molecular cations as a function of the electronic polarizability, α_0, calculated from Equations 2 and 6. The sloping line represents the trend reported in Figure 7 for alkylbenzenes. The Y-substituents are shown.

D. Mass Displacement and the Dipole Moment of the Neutral Precursor

Let us examine this situation further to determine whether the dipole moment, μ, of the neutral precursor modifies the behavior of these ions.[15] Is the departure from the sloping line (i.e., from a simple dependence on the electronic component of the polarizability) of the data points of those analytes having high permanent dipolar components related to the magnitude of the dipole moment? The difference between the polarizability, α_0, and that value which would be required to locate the corresponding data points in Figure 8 on the sloping line can be defined as $\Delta\alpha_0$. When $\Delta\alpha_0$ for the above analytes is plotted as a function of the dipole moment μ of the neutral precursor, a strikingly linear relationship is found, as is shown in Figure 9. These results constitute good proof of the effect of polarizability on mass displacements. The existence of a linear relationship between the magnitude of the dipole moment μ of a neutral species and the value for δ_m of the corresponding ionic species is strongly supportive of the hypothesis that ion dipole moments do not differ significantly from those of the corresponding neutrals. The straight line shown in Figure 9 can be considered, in principle, as a calibration curve between μ and δ_m. In order to verify this assumption, the dipole moment of $[C_6H_5\text{-}OCH_2CH_3]^+$ was calculated from the mass displacement to be 1.9 D; this value is comparable to the value of 1.45 D found for the neutral. Therefore, because dipole moments are not available for all com-

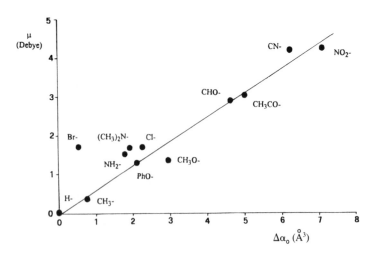

FIGURE 9

The dipole moment, μ, of neutral substituted benzene compounds plotted against the observed departure, $\Delta\alpha_0$, from the linear dependence of the mass displacement of the corresponding molecular cations upon the calculated electronic polarizability.

pounds and, in particular, for ionic species, this approach can be used easily for an evaluation of this parameter.

V. CONCLUSION

The results obtained to date show that the mass displacement with an ITMS™ instrument, as described above, can be related neither to instrumental defects nor to the occurrence of gas phase ion/molecule interactions. On the contrary, a clear relationship was established between ion structure and the observed mass displacement, which leads us to consider that the ITMS™ instrument is a unique tool for the evaluation of the polarizability of gaseous ions.

REFERENCES

1. Ion Trap Detector Manual, Finnigan MAT, San Jose, CA, 1991.
2. Kaiser, R.E.; Cooks, R.G.; Stafford, G.C.; Syka, J.E.P.; Hemberger, P.H. *Int. J. Mass Spectrom. Ion Processes.* 1991, 106, 79.
3. Traldi, P.; Curcuruto, O.; Bortolini, O. *Rapid Commun. Mass Spectrom.* 1992, 6, 410.
4. Bortolini, O.; Catinella, S.; Traldi, P. *Org. Mass Spectrom.* 1992, 27, 927.
5. Traldi, P.; Favretto, D.; Catinella, S.; Bortolini, O. *Org. Mass Spectrom.* 1993, 28, 745.
6. March, R.E.; Hughes, R.J. *Quadrupole Storage Mass Spectrometry.* Chemical Analysis Series, Vol. 102. Wiley Interscience: New York, 1989.
7. Cooks, R.G.; Kaiser, R.E. *Acc. Chem. Res.* 1990, 23, 213.
8. Todd, J.F.J.; Penman, A.D. *Int. J. Mass Spectrom. Ion Processes.* 1991, 106, 1.
9. Strife, R.J.; Kelley, P.E.; Weber-Grabau, M. *Rapid Commun. Mass Spectrom.* 1988, 2, 105.
10. Gronowska, J.; Paradisi, C.; Traldi, P.; Vettori, U. *Rapid Commun. Mass Spectrom.* 1990, 4, 313.
11. Traldi, P.; Favretto, D. *Rapid Commun. Mass Spectrom.* 1992, 6, 543.
12. Kittel, C. *Introduction to Solid State Physics.* John Wiley & Sons: New York, 1968; p. 375.
13. Partington, J.R. *Advanced Treatise of Physical Chemistry.* Vol. V. Longmans Green: London, 1961.
14. Moelwyn-Hughes, E.A. *Physical Chemistry.* 2nd ed. Pergamon Press: Oxford 1961; p. 382.
15. Bortolini, O.; Catinella, S.; Traldi. P. *Org. Mass Spectrom.* 1993; 28, 428.
16. Bortolini, O.; Olimpieri, L.; Traldi, P. *Org. Mass Spectrom.* 1994, 29, 273.

PART 4

Lasers and the Ion Trap

Chapter 5

PHOTODISSOCIATION IN THE ION TRAP

James L. Stephenson, Jr. and Richard A. Yost

CONTENTS

I. Introduction 164

II. Infrared Multiphoton Dissociation in the
Quadrupole Ion Store 166

III. UV-VIS Photodissociation in the Ion Trap Mass
Spectrometer Using a Fiberoptic Interface 169
   A. Ion Energy Deposition 170
   B. Gas Chromatography/Mass Spectrometry Coupled with
       UV Photodissociation for Isomer Differentiation 171

IV. Photodissociation in the Cylindrical Ion Trap 173
   A. Multiphoton Ionization and Trapping of $C_6H_5Cl^+$ 173
   B. Photodissociation and Spectroscopy of Trapped Ions .. 174
   C. Electronic Spectra of Complex Cations: Trapped
       Ion Photodissociation Spectroscopy 174
       1. Electronic Spectrum of the Hydrogen-Bonded
           Complex Cation $[C_6H_5OH-N(CH_3)_3]^+$ 176
       2. Stable Forms of Phenol-Complex Cations
           Using Trapped Ion Photodissociation 177

V. Ion Tomography in Ion Trap Mass Spectrometry 180
   A. Positions, Velocities, and Kinetic Energies of
       Resonantly Excited Ions 182

0-8493-4452-2/95/$0.00+$.50
© 1995 by CRC Press, Inc.

B.   Ion Frequency Determination Using a Fast Direct
     Current Pulse Pump and a Laser Probe 186

VI. Infrared Multiphoton Dissociation Studies Using a Multipass
     Optical Arrangement in Ion Trap Mass Spectrometry 192
     A.   Kinetics and Mechanism for the Infrared
          Multiphoton Dissociation of Protonated Diglyme 194
     B.   Collisional Effects for the Infrared Multiphoton
          Dissociation of Protonated Diglyme 196
     C.   Effect of Ion Storage Conditions on Photodissociation
          Efficiency 199

VII. Conclusions and Future Work 199

References 201

I. INTRODUCTION

Over the past 15 years photodissociation has become an integral tool in the study of gas-phase ion chemistry. The combination of mass spectrometry (employing both ion trap and ion cyclotron resonance instruments) and photodissociation has been used successfully to investigate the chemical kinetics, reactivity, and spectroscopy of various ionic species. The long storage times and instrumental configuration of trapping instruments are ideally suited for photodissociation experiments. Some advantages of trapping instruments include the measurement of photon-induced ion decay as a function of laser irradiance time, use of the multiphoton absorption processes to study fragmentation, and use of the photodissociation spectrum as a fingerprint for determination of isometric ion structures.

Photoinduced dissociation (PID) is the next most frequently used method for activation of polyatomic ions after collisional activation. The range of internal energies present after the photon absorption process is much narrower than those obtained with collisional energy transfer. Therefore, the usefulness of PID for the study of ion structures is greatly enhanced. However, the reduced absorption cross-sections observed with photodissociation (10^{-2} Å^2) as compared to those of collision-induced dissociation (CID; 10 to 200 Å^2) can limit this technique for analytical applications. The recent availability of higher-powered light sources over a wider range of wavelengths should provide greater flexibility for photodissociation as a routine analytical technique.[1]

Although a rigorous theoretical treatment of the photodissociation process is beyond the scope of this text, a brief summary of the basic concepts involved is presented. The process of photodissociation for a positive ion can be described by the following equation:

$$A^+ \quad \overset{nh\nu}{\underset{\text{relaxation}}{\overset{\rightarrow}{\leftarrow}}} \quad A^{+*} \quad \underset{\text{dissociation}}{\rightarrow} \quad P^+ + N \tag{1}$$

where A^+ is the ion of interest, n is the number of photons absorbed, $h\nu$ is the photon energy, A^{+*} is the excited state, and P^+ represents the product ions (with loss of neutral N). For photodissociation to occur, several prerequisites must be met. The most important criteria include the absorption of photons with energy $h\nu$, the existence of excited states above the dissociation threshold, a slow relaxation rate compared to light absorption (multiphoton processes), and dissociation rates which are fast on the time scale of the type of mass spectrometer employed.[1,2]

Several approaches for characterization of photodissociation behavior in the ion trap are discussed in this chapter. These include: (1) infrared multiphoton dissociation (IRMPD) studies of proton-bound dimers in the quadrupole ion store (QUISTOR), (2) the energetics of ion activation for *n*-butylbenzene using both photodissociation (UV-VIS) and CID, (3) single photon absorption studies as applied to isomer differentiation, (4) the use of the cylindrical ion trap (CIT) to determine dissociation thresholds of ions generated by multiphoton ionization, (5) gas phase spectroscopy of complex cations in the CIT using UV-VIS photodissociation, (6) an investigation of ion trajectories based on spatially resolved PID, (7) determination of ion frequencies using a fast direct current (DC) pulse with a laser probe, and (8) IRMPD studies using a multipass optical arrangement.

The first applications of photodissociation to trapped polyatomic organic ions were carried out in an ion cyclotron resonance (ICR) mass spectrometer by Dunbar and co-workers.[2-4] In these experiments light beams in the ultraviolet to visible region were used; the light beams were obtained from both laser (pulsed and continuous) sources and a lamp plus monochromator source. Brauman and co-workers[5] studied the vibrational relaxation of gas-phase ions and the accompanying physics of pulsed megawatt infrared multiphoton dissociation using a CO_2 laser. The ability to trap ions for an extended period of time at the low pressures achieved in the ICR cell allows for the sequential absorption of infrared photons using low intensity continuous-wave (CW) infrared radiation. This infrared multiphoton dissociation process was first characterized by Beauchamp and co-workers[6,7] for positive ions in the ICR cell. For the case

of trapped negative ions, irradiation can produce both a photodissociation and an electron photodetachment spectrum.[8-10] Several recent reviews
focusing on trapped ion (ICR) photodissociation can be found in the literature.[11,12]

II. INFRARED MULTIPHOTON DISSOCIATION IN THE QUADRUPOLE ION STORE

As with the ICR technique, the quadrupole ion trap is capable of storing ions for long periods of time. This storage capability makes the
quadrupole ion trap mass spectrometer (QITMS) very compatible with a
wide range of experiments using light. One of the first successful uses of
the ion trap in conjunction with photodissociation involved the study of
the proton-bound dimer 2-propanol utilizing a cw infrared CO_2 laser.[13] 2-
Propanol was chosen for study because its gas-phase ion chemistry is
well known, and the formation of the protonated dimer is easily accomplished. The instrumental configuration consisted of an ion trap connected
directly to the ion source of a quadrupole mass filter. The operation of
the ion trap in this manner is called a QUISTOR. The ion source located
between the ion trap and the quadrupole mass filter was used for mass
calibration purposes with perfluorokerosene. The ion optics of the mass
filter ion source allowed for transmission of ions from the ion trap to the
mass filter. The QUISTOR was operated exclusively in mode II, with the
end caps held at ground and a radiofrequency potential ϕ_0 applied to the
ring electrode. A detailed description of the optimization of ion trapping
parameters for studies of ion photodissociation in a QUISTOR can be
found in Hughes et al.[14]

These earlier studies of the IRMPD process in the ion trap were performed with a single-pass ring electrode design, with a 3-mm diameter
hole on the center axis of the ring electrode as the entrance aperture for
the low power cw CO_2 laser beam. Upon reaching the other side of the
ring electrode, a portion of the beam passed through a 0.8-mm diameter
hole and through a NaCl window, where the laser power was monitored
externally. The remainder of the laser beam was reflected by the ring electrode throughout the QUISTOR. The pressure of 2-propanol was adjusted
to 5 mPa so that photodissociation of the proton-bound dimer, $(2M + H)^+$,
at m/z 121 could occur at an appreciable rate. At pressures optimum for
the formation of the proton-bound dimer (13 mPa), no laser-induced dissociation was observed due to collisional deactivation of the vibrationally
excited proton-bound dimer. At the low pressures used in these experiments, the dissociation rate constant k_D was related to the phenomeno-
logically defined cross-section σ_D and the photon flux ϕ by the following
equation:

$$k_D = \sigma_D \phi \qquad (2)$$

The highest absorption cross-section for 2-propanol was found at 944 cm^{-1}, with the corresponding absorption of ten photons. The dissociation rate constant k_D was determined to be 2.2 s^{-1}, assuming first-order dependence on photon flux.

A typical photodissociation experiment for the proton-bound dimer of 2-propanol (m/z 121) is shown in Figure 1. The two single ion monitoring traces for m/z 121 demonstrated the mass isolation and photodissociation of the proton-bound dimer species using the QUISTOR. Photodissociation followed by subsequent mass analysis showed three different photoreaction channels open with IRMPD of the proton-bound dimer of 2-propanol, as shown below:

$$(i - C_3H_7OH)_2 H^+ + nh\nu \rightarrow (i - C_3H_7)_2 OH^+ + H_2O$$
$$m/z \; 121 \qquad\qquad\qquad m/z \; 103 \qquad (3)$$

$$(i - C_3H_7OH)_2 H^+ + n'h\nu \rightarrow (i - C_3H_7OH_2^+) + (i - C_3H_7OH)$$
$$m/z \; 121 \qquad\qquad\qquad m/z \; 61 \qquad (4)$$

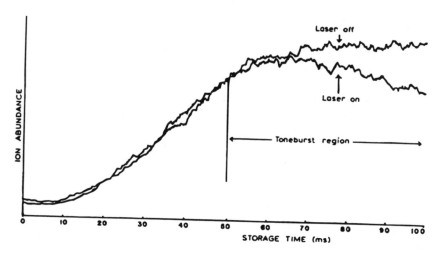

FIGURE 1

Ion abundance of [(CH$_3$)$_2$CHOH]$_2$H$^+$ vs. storage time, showing multiphoton-induced dissociation using a pulsed CW laser at 944 cm^{-1} and 80 W cm^{-2}. The upper trace shows the growth of the signal for the proton-bound dimer. At 50 ms storage time, (CH$_3$)$_2$ CHOH$_2^+$, the precursor of the proton-bound dimer, was resonantly ejected by a 206-kHz tone burst of 50 ms duration. The lower trace was obtained under conditions similar to those for the upper trace, but with additional laser irradiation beginning at 50 ms storage time and continuing for 50 ms. Ions were created at 5.3 mPa by a 1-ms, 70-eV pulsed electron beam. (Reproduced by permission of Elsevier Science Publishers B.V. from the *International Journal of Mass Spectrometry and Ion Physics*, vol. 42, 1982.)

$$(i - C_3H_7OH)_2 H^+ + n''h\nu \rightarrow (i - C_3H_7OH)H^+(OH_2) + C_3H_6 \tag{5}$$
$$m/z\ 121 \qquad\qquad\qquad m/z\ 79$$

where $nh\nu$ refers to a given number of photons absorbed. March and Hughes give a detailed description for the verification of the various reaction pathways, the photodissociation of the various isotopic analogues for 2-propanol, and the ion relaxation processes involved.[15,16]

The same experimental apparatus was used to investigate the gas-phase ion chemistry of ethanethiol, 1- and 2-propanethiol, and 1-hydroxyethanethiol.[17,18] The collisionally cooled proton-bound dimers of ethanethiol, 1-propanethiol, and 2-propanethiol were unaffected by laser irradiation at 944 cm^{-1}. However, the proton-bound dimer of 2-hydroxyethanethiol was thought to contain a S—H$^+$—O linkage which absorbs readily at 944 cm^{-1}, as opposed to the S—H$^+$—S linkage, which was shown to be transparent at the same wavelength. Isomer differentiation by multiphoton dissociation of the proton-bound dimer of propanone (m/z 117) and protonated diacetone alcohol (m/z 117) was also demonstrated.[19–21] The authors were able to differentiate the two isomeric m/z 117 ions by an aldol condensation reaction observed with the fragmentation of protonated diacetone alcohol. This reaction was not observed with the propanone dimer in the gas phase.

Investigation into the wavelength dependence of IRMPD efficiency for the proton-bound dimer of ethanol using the QUISTOR demonstrated the observable frequency shifts of the C—O stretch in the infrared region, as seen in Figure 2.[16,22] Application of the QUISTOR technique to the study of photodissociation rates by varying relaxation time, buffer gas pressure, and analyte pressure has yielded data for the proton-bound dimers of isopropanol, 2-d_1-2-propanol, and ethanol.[16] The authors also studied the effect of collision rate on the defined photoabsorption cross-section. The results obtained for the fully relaxed proton-bound dimer population showed

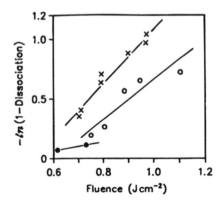

FIGURE 2
Wavelength dependence of IRMPD. Plot of $-\ln(1-I/I_0)$ of proton-bound dimers of ethanol as a function of laser fluence; (X) 944 cm^{-1}, (O) 970 cm^{-1}, (●) 1047 cm^{-1}. I was the signal intensity of the dissociating ion measured at the end of the exposure period and I_0 was the signal intensity after the same time period without irradiation. (Adapted by permission of John Wiley & Sons Publishers, *Quadrupole Storage Mass Spectrometry*, 1989.)

access to the lowest E_a pathway, thus demonstrating the only variable observed was that of the collisional deactivation process (corresponding to higher collision rates ≥ 5 ms^{-1}).

III. UV-VIS PHOTODISSOCIATION IN THE ION TRAP MASS SPECTROMETER USING A FIBEROPTIC INTERFACE

The application of a fiberoptic for UV-VIS photodissociation in the quadrupole ion trap was first demonstrated by Louris et al.[23] The instrumental design incorporated a prototype Finnigan Ion Trap Mass Spectrometer (ITMS™), with the ion trap mounted on an optical rail system, as shown in Figure 3.[24] The pulsed output of a Nd:YAG laser supplied with harmonic generating crystals was focused through a 1-mm diameter (65 cm length) PCS 1000 plastic-clad fused silica optical fiber. The fiber was then inserted through a 1.3-mm diameter hole in the ITMS ring electrode for photodissociation purposes. The population of parent ions was exposed to five laser pulses at approximately 6 to 20 mJ energy per pulse before mass analysis. The time between laser pulses was 100 ms. Typical sample pressures were 5.0×10^{-7} torr, with a helium bath gas pressure of 2.0×10^{-4} torr. The ions were illuminated with the 532 nm line (2.33 eV) of the laser. The photodissociation efficiencies observed were on the order of 5 to 30%, and in some instances under favorable condi-

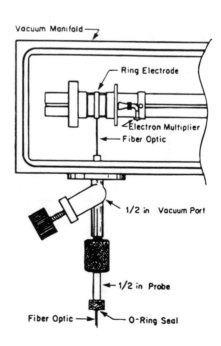

FIGURE 3

Schematic diagram of the fiberoptic interface to the ITMS. (Reproduced by permission of Elsevier Science Publishers B.V. from the *International Journal of Mass Spectrometry and Ion Processes*, volume 75, 1987.)

tions surpassed 95%. *n*-Butylbenzene was used to characterize the photodissociation process, as the energetics and fragmentation pathways are well understood. For *n*-butylbenzene photodissociation efficiency was 98% as compared to 90% observed for CID, as seen in Figure 4. The authors attributed the high photodissociation efficiencies mainly to the optical characteristics of the instrumental design, including the divergence of light from the exit of the fiberoptic, and the reflection of light from the polished stainless steel surface of the convex ring electrode, which allowed the remainder of the trap volume to be illuminated.

A. Ion Energy Deposition

The PID spectrum of n-butylbenzene in Figure 4c shows a greater ratio of $C_7H_7^+$ (*m/z* 91) to $C_7H_8^+$ (*m/z* 92) than in the corresponding CID spectrum (Figure 4b). The $C_7H_7^+$ (*m/z* 91) product ion from *n*-butylbenzene has a high activation energy and originates from the direct cleavage of the molecular ion $C_6H_5C_4H_9^+$. On the other hand, the $C_7H_8^+$ product ion possesses a low activation energy and is the result of a rearrangement process. These results follow the observed activation energies for the given fragment ion, thus demonstrating the higher energy deposition available with photodissociation. The observed ratio of *m/z* 91 to *m/z* 92 was dependent upon the number of laser pulses and the laser power.

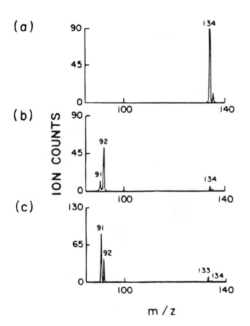

FIGURE 4

(a) Mass spectrum; (b) CID daughter ion spectrum of M⁺ (*m/z* 134); and (c) photodissociation daughter spectrum of *n*-butylbenzene in a quadrupole ion trap, power density = 10^8 W/cm², *t* = 10 ns, λ = 532 nm. (Adapted by permission of Elsevier Science Publishers B.V. from the *International Journal of Mass Spectrometry and Ion Processes*, volume 75, 1987.)

Photodissociation experiments conducted with benzaldehyde and perfluoropropylene at 355 and 266 nm, respectively, were also described. As compared to the corresponding low-energy CID data generated with the ion trap, the photodissociation data parallel high-energy CID data typically obtained with a magnetic sector tandem mass spectrometry (MS/MS) instrument.[23]

B. Gas Chromatography/Mass Spectrometry/Mass Spectrometry Coupled with UV Photodissociation for Isomer Differentiation

Using a similar experimental design, Creaser et al.[25] demonstrated the application of gas chromatography/tandem mass spectrometry (GC/MS/MS) with photodissociation. Two instrumental configurations were utilized for introduction of a fiberoptic directly into the ring electrode of the ITMS. One configuration involved the insertion of the fiberoptic through the solids probe lock, as reported previously.[23] The second configuration consisted of a modified flange with a 1/16 in. swagelock union mounted perpendicular to the probe lock and the GC interface. The output from a given harmonic of a pulsed Nd:YAG laser or a pulsed Nd:YAG pumped R6G dye laser was coupled to a 1-mm silica fiber. The beam was attenuated to 2 mJ VIS or 1 mJ UV so as to prevent damage to the fiberoptic. The transmission characteristics were >50% for all wavelengths in excess of 250 nm. The laser was pulsed using the control voltage for the auxiliary radiofrequency (RF) potential, thus disabling axial modulation during the mass analysis scan.

A mixture of *tert-*, *iso-*, *sec-*, and *n*-butylbenze was separated on a 30 m × 0.22 mm i.d. J&W DB-17 column installed into a Varian 3400 gas chromatograph. The gas chromatograph was operated isothermally at 80°C, with an injector and transfer line temperature of 250°C. The alkylbenzene mixture was ionized by electron ionization, with the molecular ions at m/z 134 RF/DC isolated. The trapped ions were exposed to 573 nm photons from the dye laser pumped with the 532 nm harmonic of the Nd:YAG laser. Photodissociation was accomplished using five laser pulses for a total scan time of 1 s. The total ion current (TIC) for the photodissociation products and the corrected CID background-subtracted spectra are shown in Figure 5.

Previous isomer differentiation studies by Uechi and Dunbar[26] successfully used the branching ratio between two competitive dissociation reactions to correlate the ratio of m/z 91 (tropylium ion) to m/z 92 (methylenecyclohexadiene ion) to a given internal energy of the *n*-butylbenzene parent ion. The ratio between m/z 91 and m/z 92 was dependent on the laser power and the number of pulses used. The product ion spectra obtained for *iso-* and *n*-butylbenzene were demonstrated to be consistent

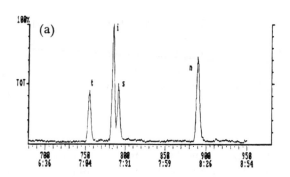

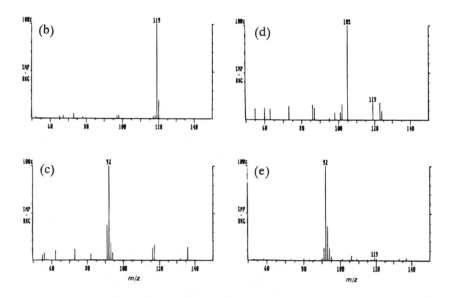

FIGURE 5

(a) Total ion current for the photodissociation products of the molecular ions (*m/z* 134) of a mixture of *tert*-, *iso*-, *sec*-, and *n*-butylbenzene separated by G-C, and the photodissociation product spectra of the molecular ions (*m/z* 134) of the mixtures of (b) *tert*-, (c) *iso*-, (d) *sec*-, and (e) *n*-butylbenzene. (Adapted by permission of Heydon & Son LTD. from *Organic Mass Spectrometry*, volume 26, 1991.)

with photons of 2.16 eV (573 nm) energy.[25] Differentiation between the *tert*- and *iso*-butylbenzene isomers was established by the presence of the *m/z* 119 base peak of *tert*-butylbenzene, indicating the loss of the methyl radical ·CH₃. The photodissociation of *sec*-butylbenzene brought about the loss of the ·C₂H₅ radical from the molecular ion of *m/z* 134 to produce the $C_6H_5CHCH_3^+$ ion at m/z 105. Typical photodissociation efficiencies were in the range of 57 to 89%.[25]

IV. PHOTODISSOCIATION IN THE CYLINDRICAL ION TRAP

The CIT has also been employed in photodissociation experiments by Mikami et al.[27-29] Using two-color laser spectroscopy in conjunction with supersonic molecular beams, ions generated by multiphoton ionization of chlorobenzene were trapped for >10 ms and were subsequently photodissociated using a second dye laser. The cylindrical electrode of the CIT cell was constructed with $r_0 = 2.0$ cm and $z_0 = 1.4$ cm. A schematic diagram of the experimental apparatus can be seen in Figure 6. One endcap of the CIT was made of stainless steel mesh for evacuation of the residual vapor after injection of the pulsed molecular beam. The mesh electrode was held at ground potential for ion injection, and also functioned as a repeller held at a potential of +10 V (1 μs) for ejection into a quadrupole mass filter. Four holes were drilled in the cylindrical electrode for the laser beams and the inlets and outlets of the molecular beams.[28] Two detailed descriptions of the CIT technique have been published.[30,31]

A. Multiphoton Ionization and Trapping of $C_6H_5Cl^+$

A pulsed nozzle system was used to generate a supersonic free expansion of gaseous C_6H_5Cl at room temperature. The resulting molecular beam was skimmed and brought into the CIT through one of the inlet holes. The beam pulse of 400 μs was photoionized at the center of the cell, with a dye laser tuned to 269.9 nm for excitation of the 0-0 band of the $S_1 \leftarrow S_0$ transition. Ion formation occurred at the crossing point of the molecular beam and photoionizing radiation. The ultracold photoionized chlorobenzene molecules were easily trapped with the CIT due to the low translational energy (tens of millielectronvolts) of the neutral molecules

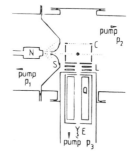

FIGURE 6
Schematic of the CIT/photoionization/photodissociation apparatus. N: pulsed nozzle, S: skimmer, C: CIT cell. L: ion lens, Q: quadrupole mass filter, E: electron multiplier. Laser beams come from the direction perpendicular to the plane. Three chambers are evacuated to $p_1 = 5 \times 10^{-4}$, $p_2 = 2 \times 10^{-5}$, $p_3 = 1 \times 10^{-5}$ torr. (Reproduced by permission of Elsevier Science Publishers B.V. from *Chemical Physics Letters*, volume 166, 1990.)

in the beam. The radiofrequency applied to the cylindrical electrode (trapping field) was between 100 and 150 kHz with amplitudes of <50 V.[28]

B. Photodissociation and Spectroscopy of Trapped Ions

Photodissociation was accomplished with the output of a Nd:YAG pumped dye laser tuned to 281.0 nm. The normal spectrum of two-photon ionized C_6H_5Cl, the mass isolated molecular ion region (M^+ and $M+2^+$) at m/z 112 and 114 (after a 2-ms storage delay), and the photolysis products of trapped $C_6H_5Cl^+$ (after a 2-ms storage delay) can be seen in Figure 7. By applying an RF potential at 130 kHz with 35 V amplitude to the cylindrical electrode, only the parent ion cluster at m/z 112 and 114 was effectively stored. Therefore, the major dissociation fragment observed in Figure 7c was due exclusively to the following reaction:

$$C_6H_5Cl^+ + hv \rightarrow C_6H_5^+ + Cl^{\cdot} \qquad (7)$$

The characteristics of the CIT cell were found to be ideal for the spectroscopic investigation of molecular ions. For the optimized trapping conditions of a given parent ion, any daughter ion generated by laser photolysis was immediately ejected from the cell without the need for perturbation of the parent ion trajectories. Studies that examined the spectrum of $C_6H_5Cl^+$ as a function of wavelength showed a steep increase in the relative yield of $C_6H_5^{+\cdot}$ from 335 to 317 nm (see Figure 8), which suggested a dissociation threshold between 3.757 and 3.910 eV. This compared well with the dissociation threshold value of 3.81 eV (indicated by the arrow in Figure 8) reported in the literature for C_6H_5Cl (3.81 eV above the ionization potential of 9.066 eV).[32] The general shape of the yield spectrum shown in Figure 8 can be explained by considering the two-photon energy of the multiphoton ionization process (energy of the ionizing light was 9.187 eV). The difference between this and the reported ionization potential of the $C_6H_5Cl^+$ is 0.121 eV. A portion of this excess energy was taken away with the ejection of an electron, leaving the remaining portion stored by the vibrationally excited ion. Even though infrared cooling can be dominant at longer storage times, the observed relaxation rate may not be fast enough to compete with the cooling process; thus, the onset of the photodissociation spectrum appears at about 0.060 eV below the reported value.[27]

C. Electronic Spectra of Complex Cations: Trapped Ion Photodissociation Spectroscopy

Spectroscopic studies of molecular cluster ions not only provide information about the structure and dynamics of molecules, but also can contribute to our understanding of the corresponding neutral clusters.

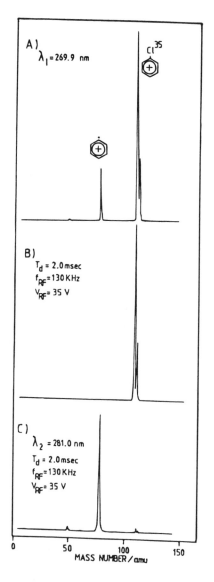

FIGURE 7
Mass spectra after the photoionization of
C_6H_5Cl. (A) Spectrum obtained with zero
delay time after the two-photon ionization
with $\lambda_1 = 269.9$ nm. (B) Spectrum obtained
by applying the pulsed repeller with a
delay time $T_d = 2$ ms from the ionization.
The frequency f_{RF} and the voltage V_{RF} of the
AC potential for the CIT cell are shown. (C)
Spectrum obtained after the photolysis of
the trapped $C_6H_5Cl^+$ by $\lambda_2 = 281.0$ nm with
$T_d = 2$ ms from the λ_1 pulse. The trapping
conditions are the same as those in (B). The
intense peak at 77 u is due to the major
daughter ion $C_6H_5^+$. The minor daughter ion
appears at 50 ± 2 u because of the
photolysis with high photon energy of λ_2
(photoionizing laser). (Reproduced by
permission of Elsevier Science Publishers
B.V. from *Chemical Physics Letters*, volume
166, 1990.)

The use of resonance-enhanced multiphoton ionization (REMPI) with the
CIT, as reported previously, can be applied successfully to spectroscopic
studies of molecular cluster ions. Thus, trapped ion photodissociation
spectroscopy, or TIP (as developed by Mikami et al.), would be a practical method for obtaining the electronic spectra of molecular cluster ions.[28]

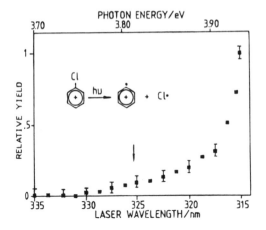

FIGURE 8
Yield spectrum of $C_6H_5^+$ from the trapped $C_6H_5Cl^+$. The yield is normalized with respect to the power spectrum of the dye laser used. The arrow indicates the reported value of the dissociation threshold. (Reproduced by permission of Elsevier Science Publishers B.V. from *Chemical Physics Letters*, volume 166, 1990.)

1. Electronic Spectrum of the Hydrogen-Bonded Complex Cation [C₆H₅OH-N (CH₃)₃]⁺

The hydrogen-bonded phenol-trimethylamine complex was prepared by a supersonic free expansion of He gas at approximately 2 atm, which contained C_6H_5OH and $N(CH_3)_3$ vapors. A 1.0-mm diameter skimmer was placed 10 mm downstream from the nozzle (orifice 0.8 mm) to skim the free jet. The pulsed molecular beam (0.7 ms duration) was ionized by REMPI using UV light obtained by SHG of a Nd:YAG laser-pumped dye laser (rhodamine 6G in methanol solution). The cluster ion was stored in the CIT for 10 ms, followed by laser irradiation in the range of 350 to 430 nm using a XeCl excimer laser-pumped dye laser with DMQ, BBQ, PBBO and Stilben-3 dyes in methanol or *p*-dioxane solutions. The photofragments were then ejected from the CIT and analyzed by a quadrupole mass filter.[28]

When the trapped cluster ion was irradiated by the excimer laser, the major product ion observed was that of the protonated amine $H^+N(CH_3)_3$ at *m/z* 60. The yield of the protonated amine (*m/z* 60) from the complex cation was proportional to the laser intensity. The TIP spectrum of the complex ion obtained by scanning the laser frequency and monitoring the formation of the protonated amine at *m/z* 54 is seen in Figure 9. The product ion yield depends on both the absorption cross-section of the complex ion and the dissociation yield in the excited state. Therefore, the TIP spectrum does not necessarily parallel the electronic absorption spectrum. Absorption peaks at 395, 379, and 365 nm seen in the TIP spectrum almost exactly match the absorption bands obtained from the resonance Raman spectrum of the phenoxy radical in solution (sharp intense peak at 399 nm, moderately intense peak at 383 nm, and a weak band at 370 nm).[33,34] The marked resemblance between the two spectra means that the chromophore of the complex ion should be the phenoxy radical. Further evidence for the phenoxy radical chromophore was the presence of two

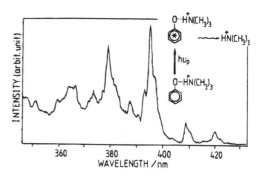

FIGURE 9
TIP spectrum of [C$_6$H$_5$OH-N(CH$_3$)$_3$]$^+$. The spectrum was normalized with respect to the laser power spectrum of the photodissociation laser. (Reproduced by permission of Elsevier Science Publishers B.V. from *Chemical Physics Letters*, volume 180, 1991.)

hot bands in the TIP spectrum at 408.5 and 420.1 nm (Figure 9). The bands at 408.5 and 420.1 nm were assigned transitions for the ground state vibrational frequencies 840 and 1520 cm^{-1}, respectively. These correspond well with the phenoxy radical vibrations observed in the resonance Raman spectrum at 841 and 1505 cm^{-1} in solution.[28]

The one-color REMPI energy of 8.81 eV used in these experiments puts a large amount of excess energy into the complex ion (actual adiabatic ionization energy of <7.0 eV). However, due to cooling processes such as collisional cooling and radiation of infrared photons occurring during the 10-ms trapping period, the vibrational temperature of the ion was found to be approximately 500 K, which was much lower than expected. Although TIP spectroscopy provides crude information (vibronic bands) about the structure of the complex ion in its stable form, extensive cooling of the trapped ions is essential for a more detailed examination of complex ion structure (rotational spectra analysis).[28]

2. Stable Forms of Phenol-Complex Cations Using Trapped Ion Photodissociation

Stable cations of phenol complexed with hydrogen-accepting molecules (NH$_3$, H$_2$O, p-dioxane) have been investigated using the TIP technique.[29] The experimental apparatus consisted of a CIT, a pumped SHG dye laser for REMPI ionization, and a XeCl excimer laser for TIP spectroscopy. The various phenol complexes were prepared in a pulsed molecular beam as described previously.[27-29]

The TIP spectrum of the phenol-NH$_3$ complex is shown in Figure 10. The spectrum was obtained by monitoring the phenol ion from the direct photolysis of the complex ion while scanning the excimer laser from 366 to 465 nm. A major band observed at 25,310 cm^{-1} was assigned to the 0-0 transition, with the remaining bands attributed to vibronic transitions. Similarities observed between the spectra of the phenol-NH$_3$ and the phenol-trimethylamine complex suggest the observed transitions were due to the phenoxy chromophore of the ion. This indicated that the phenol-

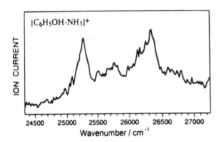

FIGURE 10
TIP spectrum of $[C_6H_5OH\text{-}NH_3]^{.}$. The $C_6H_5OH^{.}$ fragment was monitored and the ion current intensity was normalized with the power spectra of the dye lasers used. (Reproduced by permission of Elsevier Science Publishers B.V. from *Chemical Physics Letters*, volume 202, 1993.)

NH_3 complex has the proton-transferred structure of $C_6H_5O\text{-}NH_4^+$. Although this result contradicts earlier results which emphasized the nonproton-transferred structure, a reasonable explanation for this behavior was obtained by taking into account a double minimum potential surface for the complex ion, as shown in Figure 11. The authors point out the existence of two dissociation limits (shown in Figure 11) isoenergetic to one another, one which leads to the products $C_6H_5OH^{+.}$ and NH_3, and the other which yields C_6H_5O and NH_4^+. For the case in which ionization energy exceeds the dissociation limit of the complex ion, the dissociation products $C_6H_5OH^+$ and NH_3 are formed easily without the presence of a specific potential energy barrier. With the presence of a large potential energy barrier, the slow rate of proton transfer precludes the formation

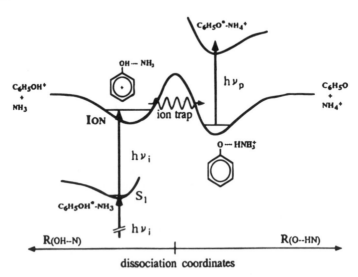

FIGURE 11
Schematic potential energy surface of $[C_6H_5OH\text{-}NH_3]^{.}$ and its excitation diagram. The dissociation limits of both channels are known to be at about the same energy. (Reproduced by permission of Elsevier Science Publishers B.V. from *Chemical Physics Letters*, volume 202, 1993.)

of the more stable proton-transferred form (C_6H_5O-NH_4^+) of the complex when the ion is excited to the dissociation continuum with a nanosecond or picosecond laser. With the use of the CIT technique (i.e., 10-ms storage period), ample time is allowed for the conversion of the metastable complex to the more stable proton-transferred form. When considering that the time scale of intracluster vibrational relaxation (IVR) is on the order of nanoseconds, the prolonged storage conditions in the CIT (millisecond time scale) allow for complete relaxation of the proton position (i.e., proton transfer dynamics) as IVR completes. Therefore, the TIP spectrum reflects the stable proton-transferred form of the complex ion, while the photoionization/dissociation results suggest the nonproton-transferred form. A detailed description of the kinetic processes involved for the $[C_6H_5OH$-$NH_3]^+$ complex ion is found in Mikami et al.[29]

The TIP spectrum of the phenol-H_2O and phenol-p-dioxane complexes were obtained from the photolysis of the complex ion to form the phenol ion and the corresponding neutral. For both complexes, broad, featureless spectra were observed which differ markedly from that obtained for the phenol-NH_3 complex, as shown in Figures 12a and b. The broadness observed with these spectra cannot be attributed to solvent interactions on the phenoxy chromophore because the absorption spectrum of the phenoxy radical in aqueous media displays a distinct vibrational structure characteristic of the chromophore. The authors report a more likely explanation for these spectra involving the phenol ion as the chromophore. Although the absorption spectrum of the phenol ion has never been pub-

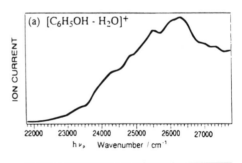

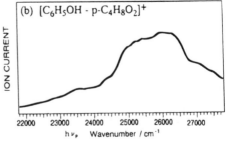

FIGURE 12
TIP spectra of (a) $[C_6H_5OH$-$H_2O]^+$ and of (b) $[C_6H_5OH$-p-dioxane$]^+$. The $C_6H_5OH^+$ fragment was monitored. (Reproduced by permission of Elsevier Science Publishers B.V. from *Chemical Physics Letters*, volume 202, 1993.)

lished, the photoelectron spectrum of phenol has an electronic excited state (2.89 eV) which corresponds well with the observed spectral region of 2.75 to 3.2 eV.[35] Previous photoelectron studies with substituted benzene cations exhibit broad, featureless spectra due to their π-π transitions.[36,37] The similarity of the photoelectron spectra to those of both the phenol-H_2O and phenol-p-dioxane complexes suggests the nonproton-transferred form as their equilibrium structure. This would account for the simple dissociation of the hydrogen bond without involving proton transfer for both the phenol-H_2O and phenol-p-dioxane complexes.[29]

V. ION TOMOGRAPHY IN ION TRAP MASS SPECTROMETRY

Hemberger et al.[38,39] developed laser photodissociation as a tool for ion tomography studies in the quadrupole ion trap. Spatially resolved photodissociation of the benzoyl cation $C_6H_5CO^+$ was used to probe ion trajectories in the ion trap as a function of the potential well depth (q_z), as a function of the applied resonant excitation voltage applied to the endcap electrodes, and with and without the presence of helium buffer gas.

The experimental apparatus, as shown in Figure 13, incorporated a modified ring electrode with a 2.0-mm wide slot cut axially through the electrode. A translatable mirror was mounted after the beam-defining optics, and just prior to entrance into the vacuum chamber. This permitted the beam to be maneuvered along the length of the slot (z-axis) in the ring electrode. The photons employed in these experiments were from an XeCl excimer laser at 308 nm with a pulse width of approximately 15 ns. The total energy observed was ≤ 100 mJ, with energies in the trap measured at 1 to 3 mJ. The beam dimensions ($1/e^2$) were 1.0 mm (Gaussian) $\times$ 3.0 mm (flat top), with the long axis perpendicular to the slotted ring electrode, and thus attenu-

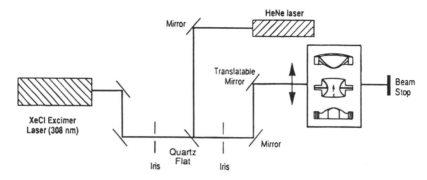

FIGURE 13

Schematic of the laser/ion trap system used for spatially resolved photodissociation studies. (Reproduced by permission of Elsevier Science Publishers B.V. from *Chemical Physics Letters*, volume 191, 1992.)

ated by the 2.0-mm wide slot. Typical sample pressures were in the 8×10^{-8} torr range, with a helium bath gas pressure of 1×10^{-5} torr. Electron ionization was used in conjunction with RF/DC isolation of the resulting m/z 105 $C_6H_5CO^+$ ion. This was followed by a 10-ms relaxation period and a single-pulse of laser irradiation with resulting mass scan from m/z 70 to 120.

The radially averaged axial distribution of benzoyl cations from $z = 0$ was approximately 1.1 mm at full width half medium (FWHM), agreeing well with the theoretically predicted distribution found in the Finnigan ion trap.[40] Experimental evidence of the collisionally stabilized orbits of ions confined to the center of the trap employing helium buffer gas also was demonstrated, as seen in Figure 14. Detection efficiency improved with the use of helium buffer gas due to the more tightly clustered ions in the center of the trap. As the voltage on the RF ring electrode, and thus q_z, was increased, the depth of the potential well for ion storage above the low mass cutoff also increased. Benzoyl ions were trapped at several q_z values, with a corresponding increase in storage efficiency observed from $q_z = 0.2$ to $= 0.6$. Calculated potential well depths for the m/z 105 ion at q_z values of 0.2, 0.4, and 0.6 were 6.5, 27, and 60 eV, respectively.

Also, an investigation of the axial ($\beta_z = 0.30$) frequency involved in resonant excitation of the benzoyl cation (m/z 105) stored at a $q_z = 0.4$ was undertaken. By applying a 164-kHz low amplitude (0.1 V_{p-p}) signal to the end-cap electrodes, the benzoyl ion trajectories were perturbed sufficiently to expand the ion cloud from the center of the ion trap, but not enough to gain ample kinetic energy to induce CID. The expansion of the ion cloud in the z direction with the application of the resonant excitation voltage was approximately double that seen without the application of a resonant excitation voltage.

FIGURE 14
The axial distribution of benzoyl ions in the trap with and without He buffer gas; ($\bullet$) no He, $t_{ion} = 6$ ms, ($\bigcirc$) He, $t_{ion} = 3$ ms. The stabilization of the distribution with He buffer gas provides enhanced mass resolution and sensitivity. (Reproduced by permission of Elsevier Science Publishers B.V. from *Chemical Physics Letters*, volume 191, 1992.)

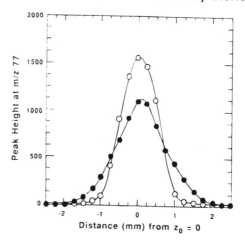

Helium Stabilization of Ion Trajectories

A. Positions, Velocities, and Kinetic Energies of Resonantly Excited Ions

Williams et al.[38,39] employed spatially resolved photodissociation (ion tomography, as described previously) of the benzoyl cation to determine the kinetic energies of resonantly excited ions in the quadrupole ion trap. The determination of ion kinetic energies in the quadrupole ion trap represents an important step in understanding the collision process (MS/MS) for analytical applications. Kinetic energy data were obtained by firing a 308-nm excimer laser at selected phase angles of the resonant alternating current (AC); ion motion were tracked in increments of <1 μs. The resulting ion displacement vs. time data were then converted to ion velocities. Variations of the ion distribution as a function of the AC phase angle (q_z = 0.2) was also described using spatially resolved photodissociation.

The instrumentation was described previously in this chapter.[38] The ion trap electronics were modified to produce a master trigger pulse at the first zero crossing of the supplementary AC voltage. The errors associated with detection of the crossing point (the main clock rate is 3.58 MHz for the Scan Acquisition Processor) were ±140 ns. This master pulse then triggered a variable delay generator (Stanford Research DG 535) to produce a slave pulse which controlled the triggering of the laser. This trigger delay allowed the laser to be fired at any given phase angle of the supplementary AC. The signal output from a photodiode and the supplementary AC voltage were sent to a 100-MHz digital oscilloscope to measure the timing of the photodissociation event with respect to the actual phase angle of the supplementary AC.[39]

A 15-ns pulse from a XeCl excimer laser (Questek model 2200) at 308 nm was used for photodissociation. The laser beam width (at the focal point inside the trap) was measured by a photodiode array positioned on the beam line external to the vacuum chamber. The beam distribution was found to be Gaussian along the z-axis, with a corresponding beam width (FWHM) of 0.64 nm. Spatial distribution measurements were obtained by monitoring the photodissociation product $C_6H_5^+$ (m/z 77) from the benzoyl ion (m/z 105) of acetophenone (mass isolation of m/z 105 from the electron ionization of acetophenone) as a function of laser beam position. The beam position was adjusted along the z-axis of the ring electrode using a mirror mounted to a translatable stage, as described previously.[38] The supplementary AC voltage applied to the end-caps resonantly excited the benzoyl ions for 10 ms. The laser was fired 7 ms into the excitation period, which allowed the ions to reach their steady-state displacements. The amplitude of the supplementary AC voltage was adjusted so that the number of $C_6H_5^+$ produced by CID was limited to <10% of the benzoyl (m/z 105) ion from acetophenone.[39]

Figure 15a shows the axial distribution of benzoyl ions centered at z = 0. The shape of this distribution resembles closely previous results reported for ions which were cooled to the center of the trap. Figure 15b, the benzoyl ion distribution, is seen as the laser was fired 7 ms after a 40-mV$_{0-p}$, 160.5 kHz resonant AC signal was applied to the end-caps. As shown in this figure, ion motion was altered significantly by application of the resonant AC voltage. Data obtained for the experiment in Figure 15b were not phase locked to the supplementary AC. This experiment was then repeated with the laser phase locked at 90°, 180°, 270°, and 350° with respect to the AC signal. The results produced in Figure 15c eliminated the uncertainties from sampling data over a large range of phase

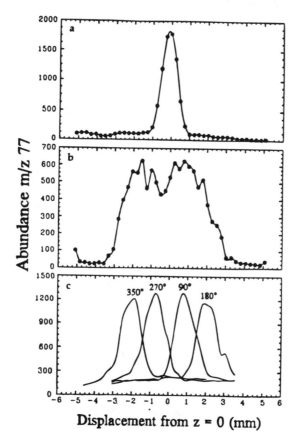

FIGURE 15

Distributions produced by spatially resolved photodissociation of benzoyl ions (a) prior to the application of the supplementary AC voltage, (b) without phase locking of the laser to the supplementary AC voltage, and (c) with phase locking at specified phase angles of the AC excitation voltage. For b and c, photodissociation was performed 7 ms into the resonance excitation period to obtain the ion distribution under equilibrium conditions. (Reproduced by permission of Elsevier Science Publishers from *Journal of the American Society for Mass Spectrometry*, volume 4, 1993.)

angles. This phase-locked experiment demonstrates that ions move as a function of the AC phase angle, and that He buffer gas does not destroy the coherence of the ion packet. Because the buffer gas acts essentially as a viscous drag, the system can be modeled as a forced damped oscillator in which damping is provided by the He buffer gas and the supplementary AC voltage is the forcing function.[41,42]

When the mean displacements of the ion distributions were plotted as a function of the applied phase angle, (as seen in Figure 16, for q_z = 0.2 and 0.4), the data points were found to fit a sinusoidal function (z = $z_{max}\sin(2\pi\omega_z + \phi)$), where z_{max} was the fitted maximum displacement, ω_z was the fundamental frequency (axial) of the ions, and ϕ was the fitted phase shift. A simple harmonic oscillator model was used to predict this function.[16] For the forced damped oscillator model, the output (ion motion) and input (applied AC) will approximate a harmonic oscillation at the same frequency. The ion motion was expected to lag the applied AC when the input frequency ω matched the resonant frequency ω_z of the system. A phase shift smaller than 90° was expected when $\omega < \omega_z$, and a phase shift larger than 90° was expected when $\omega > \omega_z$.[42] The authors re-

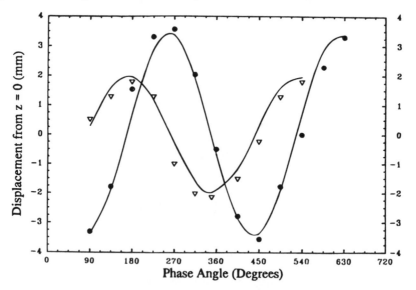

FIGURE 16
Displacements of the ion distributions (measured at the maxima of the individual distributions) as a function of phase angle at q_z values of 0.2 and 0.4. Positive displacement is for ion cloud movement toward the filament end-cap. Phase angles are relative to the AC signal applied to the exit end-cap. ● = q_z = 0.2, full line is a plot z (mm) = 3.48 (mm) × (sin2 $\pi\omega_z t$ − 169°); ∇ = q_z = 0.4; full line is a plot z(mm) = 1.97 (mm) × (sin2$\pi\omega_z t$ − 77°). (Reproduced by permission of Elsevier Science Publishers from *Journal of the American Society for Mass Spectrometry*, volume 4, 1993.)

ported a phase shift for ion motion of 169° for q_z = 0.2 and 77° for q_z = 0.4 with respect to the applied (AC) resonant excitation voltage. The magnitude of the reported phase shift values varies with AC frequency and amplitude, He buffer gas pressure, and trapping potential. Also, higher-order field effects from the "stretched" geometry of the ion trap may cause additional phase shift phenomena.

By differentiating the ion displacements with respect to time (v = dz/dt), the ion velocities were obtained for q_z = 0.2 and 0.4. A comparison of the experimental data to the fitted results (v = $2\pi\omega_z z_{max} \cos (2\pi\omega_z + \phi)$) is shown in Figure 17. The observed deviations of the fitted results (simple harmonic motion model) from the experimental data were due to the differentiation of the poorly fitted displacement points (Figure 16), which gave an even worse fit to calculated ion velocities. As the ion cloud passes through zero axial displacement, maximum ion velocities and kinetic energies were obtained. In Figure 17 this corresponds to phase angles of 169° and 349° for q_z = 0.2, and phase angles of 257° and 437° for q_z = 0.4. Therefore, the majority of activating collisions that lead to dissociation and subsequent storage of product ions are expected to occur in this region due to increased ion velocities. The axial component of ion kinetic energies in Figure 17 was <5 eV for all phase angles, which was expected due to the small amount of CID observed. By increasing the res-

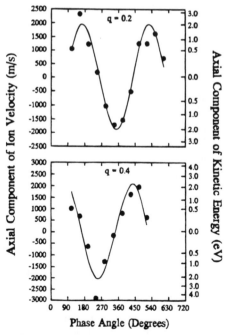

FIGURE 17
Axial component of ion velocities and kinetic energies calculated for the data shown in Figure 16. (Reproduced by permission of Elsevier Science Publishers from *Journal of the American Society for Mass Spectrometry*, volume 4, 1993.)

• Experimental Data   — Simple Harmonic Motion

onant AC voltage so that efficient CID may occur (80 mV$_{0-p}$), spatially re-solved photodissociation was used to determine the axial component of kinetic energy for the benzoyl ions. A maximum kinetic energy value of 9 eV was obtained for m/z 105 under these conditions.

The authors also reported that the width of the ion distribution changed as a function of phase angle. The ion cloud was shown to expand at the zero displacement region and contract in the maximum excursion region. This behavior was attributed to velocity compaction of the ions at z_{max} where a reduction of the ion velocity is observed as the ion cloud approaches z_{max}. The general trend observed showed a decrease in the FWHM of the ion distributions in each AC cycle, as shown in Figure 18. Calculated timing errors at the zero crossing region and at z_{max} were ±0.266 and ±0.010 mm, respectively. Although these errors distort the FWHM of the ion distributions, the general trend of cyclic variations in FWHM with phase angle was easily observed.

B. Ion Frequency Determination Using a Fast Direct Current Pulse Pump and a Laser Probe

Determining the frequency of stored ions in the radial (r) and the axial (z) directions contributes much to understanding the processes of

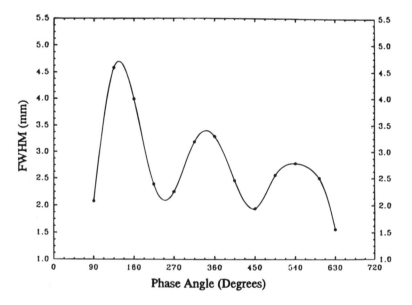

FIGURE 18
Width (FWHM) of the ion distribution plotted as a function of phase angle for a q_z value of 0.2. (Reproduced by permission of Elsevier Science Publishers from *J. Am. Soc. Mass Spectrom.*, volume 4, 1993.)

resonance excitation, mass range extension, and RF-only isolation in the quadrupole ion trap. In addition to the fundamental secular frequencies of motion for a given ion population in both the radial and axial directions, higher-order fields (i.e., hexapole and octopole) introduce additional component frequencies, the relative magnitudes of which contribute to the overall trajectory of the ions. Lammert et al.[43] developed a technique for the determination of an ion's unique set of frequencies by using a fast DC pump and a laser probe. This novel technique employs a fast DC pulse to displace a kinetically cooled ion population from the center of the trap. This electrostatic pulse causes the ion cloud to oscillate within the ion trap. As the ions reappear at the center of the trap, the ion population was probed with a laser by generating and then measuring the intensity of the photodissociation product ion. By monitoring the photodissociation products as a function of the delay time (50-ns intervals) between the DC pump and the laser probe, the time-based data obtained yield the individual ion frequencies and their relative magnitudes via a fast Fourier transform analysis. Ions were translated axially by applying the DC pulse to one end-cap while the other was grounded. Radial translation was obtained by applying the DC pulse simultaneously to both end-caps, although neither technique was found to activate the ions exclusively in one direction.

All experiments were performed using a prototype ITMS (described previously).[24] Acetophenone and 1,4-cyclohexadiene were used for all experiments because the photodissociation of the given mass-selected ion results in a single product ion with high efficiency. For acetophenone the benzoyl cation ($C_6H_5CO^+$) at m/z 105 was mass isolated and the product from photodissociation at m/z 77 ($C_6H_5^+$) was monitored. However, at q_z values >0.67 for the m/z 105 parent ion (benzoyl cation), the product ion monitored at m/z 77 has a q_z >0.908, and thus lies outside the stability boundary. The molecular ion of 1,4-cyclohexadiene ($C_6H_8^{+\cdot}$) at m/z 80 was used for studying higher q_z values because the major product of photodissociation was m/z 79 ($C_6H_7^+$). The high-voltage DC pulser applied a low amplitude (10 to 50 V for axial translation, 100 V for radial translation) 2-μs DC pulse to one or both of the end-cap electrodes for displacement of the ion population from the center of the trap.[44] All pulse initiation sequences were started at zero RF phase angle. The laser employed was a XeCl excimer at 308 nm. The 1-mm beam was steered through a 1.3-mm diameter aperture on the center axis of the ring electrode. Electronics were modified to provide a TTL pulse (phase locked to the RF at zero phase angle) for control of a pulse delay generator which precisely triggered both the DC voltage and the laser. A 1-ms delay was employed after the 2-μs DC pulse (just prior to the firing of the laser), so the ions could achieve stable orbits. The delay was then increased from 1 ms at 50-ns intervals until a sufficient number of time increments were sampled.[43]

The time and frequency domain data for $q_z = 0.3$ are shown in Figure 19. In this experiment a 10-V DC pulse was applied to one end-cap for axial activation. The time scale on the x-axis of Figure 19 refers to the delay time after a 1-ms cool time. The authors point out that the periodic photodissociation signal obtained in the pump/probe experiment (DC pulse on) maximized at the same signal level as the control experiment (DC pulse off). These results showed that the ion population was kinetically cooled within 1 mm of the center of the trap, as previously reported by Hemberger et al.[38] The large peaks observed in the time domain spectrum were due to the sampling method of the technique. Because an ion of any given frequency can cross through the sampled volume in the center of the trap twice during a complete cycle, the measured frequency would then be twice the actual frequency ($2f_z$). This behavior was expected because the complex Lissajous motion of trapped ions allowed some ions to be sampled twice per cycle. These results were confirmed experimentally by moving the laser off center with an observed increase in f_z, the fundamental secular frequency.

The agreement between the experimental $2f_z$ value of 244 kHz is within ±5% of the predicted value $2f_z = 238$ khz. Also shown in the frequency spectrum are the axial harmonics $3f_z$ and $4f_z$. These harmonics may arise from fast Fourier transform or sampling effects, the presence of nonlinear fields in the nonideal trap, or in the case of the $4f_z$ peak, sampling the second harmonic twice per cycle.[45] In addition, sideband frequencies due to the RF drive frequency (f_{RF} hexapole field contribution) and multiples of the drive frequency ($f_{RF}-f_z$, $f_{RF} + f_z$ quadrupole field contribution) were also present.

As the depth of the pseudopotential well is increased (increase in q_z), the amplitude of the DC pulse must also increase to see frequencies associated with hexapolar and octopolar fields. When a 25-V DC pulse was used at a q_z value of 0.6, a more complex time domain spectrum was obtained (Figure 20). The measured frequency of the $2f_z$ at 503 kHz agreed well with the theoretically predicted value of 508 kHz. Also present were the RF drive frequency and corresponding sideband frequencies at $f_{RF}-f_z$ and $f_{RF} + f_z$. The frequencies observed at $f_{RF} \pm 2f_z$ and $f_{RF} \pm 3f_z$ could result from nonlinear resonance fields. At q_z values >0.79 hexapolar field effects (f_{RF}) were found to be the main frequency component of ion motion. From these experiments the authors report that the stretched geometry of the Finnigan ion trap introduces additional nonlinear field effects on ion motion. The presence of nonlinear octopolar fields can introduce a small degree of radial activation. Therefore, the f_z peak observed at 242 kHz could also be due to $2f_r$, which has a calculated frequency of 238 kHz. The authors note that when the laser was sampling a cross-section not in the center of the trap in the axial direction, the maximum signal for the pump/probe experiment exceeded that of the control experiment (DC

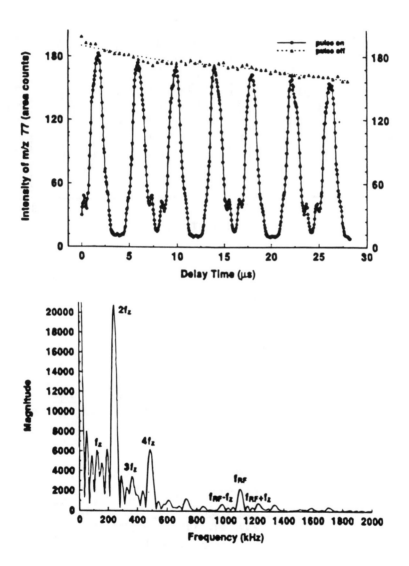

FIGURE 19
The time domain (top) and frequency domain (bottom) data for the DC pulse pump/probe experiment using a 10-V, 2-μs axial DC pulse at a q_z value of 0.3. The trapped ion is the benzoyl cation, m/z 105, and the photofragment is m/z 77. The time delay refers to the length of time between the DC pulse and the firing of the laser after allowing a 1-ms cooling time. (Reproduced by permission of Elsevier Science Publishers from *J. Am. Soc. Mass Spectrom.*, volume 5, 1994.)

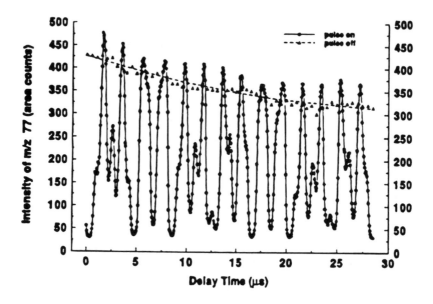

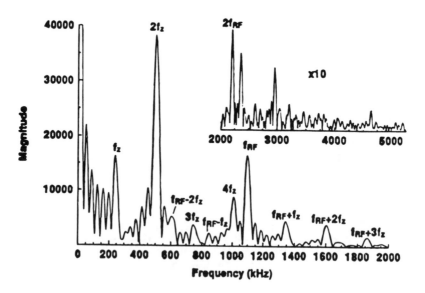

FIGURE 20

The time domain (top) and frequency domain (bottom) data for the DC pulse pump/probe experiment using a 25-V, 2-μs axial DC pulse at a q_z value of 0.6. The trapped ion is the benzoyl cation, m/z 105, and the photofragment is m/z 77. The time delay refers to the length of time between the DC pulse and the firing of the laser after allowing a 1-ms cooling time. (Reproduced by permission of Elsevier Science Publishers from *J. Am. Soc. Mass Spectrom.*, volume 5, 1994.)

pulse off). Improved performance of the experiments was obtained for off-center sampling.

Activation in the radial direction was accomplished by applying a 100-V DC pulse to both end-caps simultaneously. For a q_r value of 0.25, only the radial frequency $2f_r$ and corresponding harmonics nf_r were observed. No axial frequencies were reported for this experiment due to the reduced sensitivity to axial motion for a radial excitation experiment. Also, the sensitivity for radial (as well as axial) activation was significantly reduced because the beam only intersects a small portion of the radial plane of the ring electrode. This led to much broader frequency peaks for the radial activation experiment.

These pump/probe experiments were also simulated using the ITSIM program, a PC-based ion trap simulation program which calculates ion trajectories in both the axial and radial directions.[46] Figure 21 shows the results of an ITSIM simulation of the pump probe experiment with an axial DC pulse at 25 V and q_z value of 0.6. The phase coherence of the ions can be seen for the plot of z-position vs. time, with the general pattern of the time-domain spectrum corresponding well with the experimental data shown in Figure 20. The value of $2f_z$ from the simulation (490 kHz) agreed well ($\pm5\%$) with that obtained experimentally (503 kHz). The shoulder peaks on the simulated data, which corresponded to a frequency of 122 kHz, may be due to either radial motion (f_r) at a frequency of 119 kHz, or the possibility of a $f_z/2$ frequency at 127 kHz.

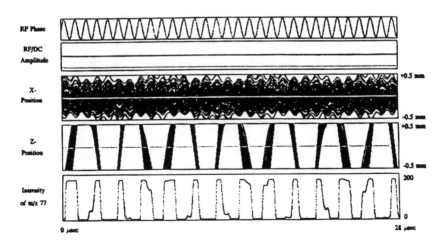

FIGURE 21

The simulated time domain data for the DC pulse pump/laser probe experiment using a 25-V, 2μs axial DC pulse at a q_z value of 0.6. (Reproduced by permission of Elsevier Science Publishers from *J. Am. Soc. for Mass Spectrom.* volume 5, 1994.)

VI. INFRARED MULTIPHOTON DISSOCIATION STUDIES USING A MULTIPASS OPTICAL ARRANGEMENT IN AN ION TRAP MASS SPECTROMETER

Infrared multiphoton dissociation was also used in conjunction with a Finnigan MAT ITMS™ quadrupole ion trap. The design of a novel multipass optical arrangement circumvents previous problems of limited infrared laser power, and small infrared absorption cross-sections for many organic molecules. The novel ring electrode in conjunction with a pulsed valve (mounted inside the vacuum chamber) for the introduction of He buffer gas greatly improves the trapping and detection of ions for IRMPD experiments in the quadrupole ion trap.[47,48] This multipass optical arrangement was constructed to produce eight laser passes across the radial plane of the ion trap ring electrode. This modified optical arrangement, as originally described by White,[49] increased the optical path length of photon absorption in the infrared region. Incorporation of this novel White-type cell design into an ICR cell was first demonstrated by Watson et al.[50] The White-type ICR cell was used for the study of resonance-enhanced two-laser IRMPD of gaseous perfluoropropene cations, gallium hexafluoroacetylacetonate anions, protonated diglyme cations, and allyl bromide cations. The authors reported that fragmentation in the White-type cell (using energies lower by almost a factor of 20) was comparable to that seen in a simple double-pass experiment.[50,51]

A 1.0-cm diameter infrared laser beam was attenuated by the 0.3 cm ($\frac{1}{8}$ in.) entrance aperture on the ring electrode. The ring electrode was modified by incorporation of three polished stainless steel spherical concave mirrors (radius of curvature, 2.0 cm) mounted on the inner surface of the ring, as shown in Figure 22. The approximate photon density (assuming constant intensity across the attenuated beam width) observed in the radial plane of the ring electrode can be seen in this figure. The most critical adjustment of the mirror system was the separation of the centers of curvature of the mirrors A and B, shown in Figure 22. This separation determines the number of beam transversals across the ring electrode: 4, 8, 12, or any other multiple of 4. The mirrors were mounted on the ring electrode in such a way that the centers of curvature of mirrors A and B were on the front surface of mirror C, and the center of curvature of mirror C was halfway between mirrors A and B.

A schematic diagram of the instrumentation employed can be seen in Figure 23. Except for the pulsed valve experiments, the ion trap was operated without He buffer gas. The software used was modified to accommodate two TTL pulses for computer control of both the cw CO_2 laser and the pulsed-valve apparatus. The cw CO_2 laser beam was reflected off a gold-plated turning mirror through a ZnSe window into the Finnigan ITMS.™ To determine the laser power entering the 0.3 cm aperture in the ring electrode, the analyzer assembly was removed from the vacuum chamber (due to space considerations) and a power meter was placed in the beam path (Coherent Radiation Model 410). Wavelength determinations were made by an infrared spectrum analyzer (Optical Engineering, Model 16-A).

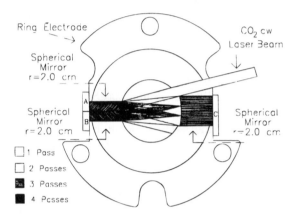

FIGURE 22
Modified ring electrode used for multipass experiments in the quadrupole ion trap. The labels A and B indicate that the mirrors adjusted to obtain eight laser passes across the ring electrode. The shaded areas indicate the approximate photon density in the radial plane of the ring electrode. (Reproduced by permission of Elsevier Science Publishers from *J. Am. Soc. Mass Spectrom.*, vol. 5, 1994.)

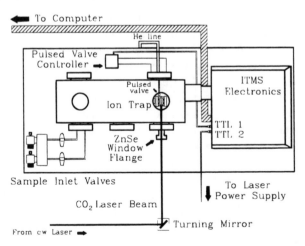

FIGURE 23
IRMPD instrumental configuration showing computer control lines, pulsed valve apparatus, and cw CO_2 laser. (Reproduced by permission of Elsevier Science Publishers from *J. Am. Soc. Mass Spectrom.*, vol. 5, 1994.)

A. Kinetics and Mechanism for the Infrared Multiphoton Dissociation of Protonated Diglyme

In a typical IRMPD experiment a 15-ms ionization period was followed by a reaction time of 400 ms, and the resulting protonated diglyme was isolated using frequency sweep, apex, or two-step isolation routines. A laser delay time of 400 ms was included to allow for removal of any excess internal energy by radiative and/or collisional cooling. Depending on the experiment, the laser irradiation time varied from 5 to 250 ms, which was then followed by periods for laser discharge and mass analysis.[47]

The increased photon absorption cross-section obtained with this multipass arrangement gave first-order rate constants larger by a factor of 50 to 100 than those with traditional single-pass designs. For example, values for the photodissociation rate constant of protonated bis(2-methoxyethyl)ether (diglyme) were typically in the 100 s^{-1} range, while those with single-pass arrangements were approximately 2 to 5 s^{-1}.

Characteristic photodissociation spectra of the $(M + H)^+$ ion of diglyme at 944 cm^{-1} are shown in Figure 24. At lower irradiation energies (irradiation time, 10 ms), the single reaction channel observed was the formation of m/z 103 with corresponding loss of neutral methanol, shown in Figure 24b. At higher irradiation energies (irradiation time, 40 ms), the dominant reaction channel involved the loss of an acetaldehyde neutral from the m/z 103 ion shown in Figure 24c. The presence of a small m/z 59 peak at lower irradiance times suggested a competitive reaction mechanism for the formation of the product ion species from the m/z 135 parent ion. Alternatively, when ion intensity was plotted as a function of laser irradiance time (Figure 25), the appearance was that of a series of consecutive reactions. To determine if the reaction mechanism was competitive or consecutive, a series of MS^n experiments was conducted. In the first experiment the formation and mass isolation of protonated diglyme (m/z 135) were achieved as described earlier. AMS/MS experiment was then performed with a 10-ms laser irradiance pulse followed by mass isolation of the m/z 103 product ion. A series of MS^3 experiments on the m/z 103 ion by varying the laser irradiances times from 1 to 80 ms (increased energy input) produced the ion growth curve for the m/z 59 product ion shown in Figure 25. To verify the exclusive formation of the m/z 59 product ion from the m/z 103 parent, a second MS/MS experiment was performed on the protonated diglyme parent ion (m/z 135) with concurrent notch-filter ejection of the m/z 103 product ion. As before, the laser irradiance time was varied from 1 to 80 ms. Results from this experiment produced only a decrease in the protonated diglyme parent ion, but no formation of the m/z 59 product ion. These experiments confirmed the formation of the m/z 59 product ion from the m/z 103 precursor, thus verifying the presence of a consecutive reaction mechanism of the type m/z 135→m/z 103→m/z 59.

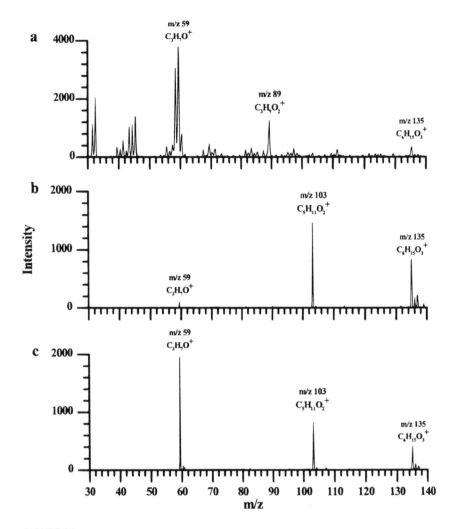

FIGURE 24

(a) Electron ionization mass spectrum of diglyme at a pressure of 3.6×10^{-7} torr and 1.5 ms ionization time (no He present). (b) IRMPD spectrum of protonated diglyme at 944 cm^{-1} and 10 ms irradiance time (energy, 0.201 J). (c) IRMPD spectrum of protonated diglyme at 944 cm^{-1} and 40 ms irradiance time (energy, 0.852 J). (Reproduced by permission of Elsevier Science Publishers from *J. Am. Soc. Mass Spectrom.*, vol. 5, 1995.)

The proposed mechanism of this consecutive reaction, seen in Figure 26, corresponds well with results reported previously for the IRMPD of protonated diglyme in the ICR cell.[50]

The minimum number of photons (n) needed to reach the thermodynamic threshold for dissociation via a given reaction channel was calculated by dividing the enthalpy change by the photon energy. The re-

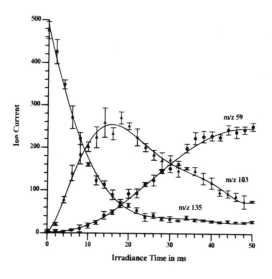

FIGURE 25
IRMPD ion growth curves for protonated diglyme as a function of laser irradiance time. Two reaction channels were observed: one for the formation of the m/z 103 fragment from the $(M + H)^+$ ion and the other for the formation of the m/z 59 fragment from m/z 103. Some backreaction of m/z 59 with neutral diglyme to reform the $(m/z$ 135) $(M + H)^+$ ion also was observed at longer irradiance times. (Reproduced with permission from Elsevier Science Publishers from *J. Am. Soc. Mass Spectrom.*, vol. 5, 1994.)

sults obtained for the reaction channels observed were $n \geq 3$ for the lower energy process, and $n \geq 9$ for the higher energy process. The calculated photon energy at the wavelength used (10.6 μm) was 0.117 eV.[48]

B. Collisional Effects for the Infrared Multiphoton Dissociation of Protonated Diglyme

Collisional deactivation studies of polyatomic species typically provide energy transfer information on a broad scale. These gross changes in energy levels are represented by changes in observed fragmentation or dissociation of given polyatomic species. Previous studies of collisional quenching (pulsed CO_2 lasers) carried out in the ICR have examined the efficiency of the quenching process.[6,52-54] For the case of low-power cw CO_2 lasers, collisional deactivation successfully competes with photodissociation at sufficiently low photon fluxes and relatively high neutral pressures.[55] Therefore, the amount of photodissociation observed relates directly to the collisional deactivation efficiency of the neutral buffer gas.

In Figure 27 the effect of collisions on the IRMPD of diglyme is shown. The general trend observed for all buffer gases showed a decrease in the photodissociation efficiency with increasing gas pressure. The observed trend for quenching efficiency was $N_2 > Ar > He$. These results correlated well with previous studies on polyatomic and diatomic ions using the same buffer gases.[55-58] Traditionally, collisional deactivation efficiency increases with mass and polarizability of the neutral molecule. For collisions involving He and Ar, the loss of vibrational energy from the pro-

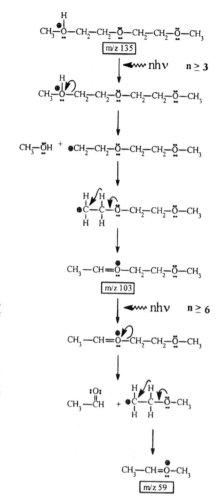

FIGURE 26
Reaction mechanism for the IRMPD of protonated diglyme using a cw CO_2 laser. The low-energy reaction channel corresponds to the absorption of a minimum of three photons (e.g., formation of m/z 103), while the high-energy reaction channel corresponds to the absorption of at least an additional 6 photons, for a total of 9 photons absorbed by the protonated species. All reaction mechanisms were confirmed by MS^n experiments. (Reproduced with permission of Elsevier Science Publishers from *J. Am. Soc. Mass Spectrom.*, vol. 5, 1994.)

tonated diglyme can occur only by a vibrational to translational energy transfer. For the N_2 collision partner, vibrational energy transfer is possible but not realistic due to the large difference in vibrational frequencies. In all probability, the greater collisional quenching efficiency of N_2 is due to its greater polarizability as compared to that of Ar. The collisional deactivation process was most efficient when the protonated diglyme molecule collided with a diglyme neutral. The close match between the vibrational frequencies of these two colliding molecules can facilitate intermolecular vibrational energy transfer.[7]

Pulsed valve experiments were conducted to determine if the increase in trapping efficiency during ionization associated with the addition of He buffer gas to the ion trap analyzer would interfere with the collision-

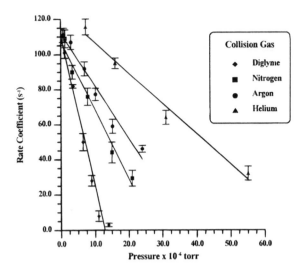

FIGURE 27
Effect of collisions on the photodissociation rate of protonated diglyme using N_2, Ar, He, and neutral diglyme as target gases. All pressure measurements are corrected values.

free requirements for IRMPD. A 1.6-ms pulse of He gas was found to trap the maximum number of diglyme $(M + H)^+$ ions for a 10 psi He back-pressure. The signal intensity of protonated diglyme (m/z 135) using a 1.6-ms pulse of He gas during the preionization period was higher by a factor of 7 than when He was not used in conjunction with the ionization event. The ionization time for both experiments was 1.0 ms.

In order for photodissociation to occur, the rate of photon absorption must be greater than that of collisional and radiative deactivation. Thus, for pulsed valve experiments, long ion storage times (several seconds) were required before triggering the laser to pump away the He buffer gas initially used to efficiently trap and damp the diglyme ions needed for the formation of the protonated (m/z 135) species. Storage efficiency measurements show an approximate loss of 2% of the original ion signal after 70 s of ion storage. For these experiments, the efficiency of photodissociation (P_D) was defined as the fraction of the initial ion population photodissociated at a designated exposure time for a given laser irradiance:

$$P_D = 1 - \frac{I}{I_0} \qquad (8)$$

where I_0 is the ion signal without irradiation at the end of the designated exposure time, and I is the ion signal of the photodissociated ion at the end of the same time period. Figure 28 shows the photodissociation effi-

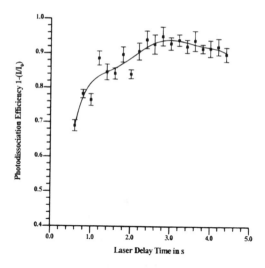

FIGURE 28
Photodissociation efficiency as a function of postpulse valve laser delay. Photodissociation efficiency remains constant at over 90% for delay times >2s. (Reproduced with permission of Elsevier Science Publishers from *J. Am. Soc. Mass Spectrom.*, vol. 5, 1994.)

ciency as a function of the laser delay time. After a 2-s delay between the He gas pulse and laser trigger, the photodissociation efficiency levels off at approximately 90%.

C. Effect of Ion Storage Conditions on Photodissociation Efficiency

The multipass optical design used in these experiments produces a high photon density in the radial plane of the ring electrode, as shown by Figure 22. A preliminary investigation to determine the influence of storage conditions (q_z and a_z) on photodissociation efficiency was undertaken with protonated diglyme. The effect on the photodissociation efficiency of exciting the $(M + H)^+$ ions of diglyme (no He present) using resonant excitation is seen in Figure 29. With increasing resonant excitation voltage applied to the end-caps, the axial excursions of the ions increase; thus, the fraction of time the ions spend in the radial plane of the multipass ring electrode, and therefore the extent of interaction between the stored ions and photons, decrease. This leads to the decrease in photodissociation efficiency observed for increasing resonant excitation voltage.

VII. CONCLUSIONS AND FUTURE WORK

Research performed with the quadrupole ion trap in conjunction with photodissociation has focused primarily on fundamental gas-phase ion chemistry, including kinetic investigations, fundamental spectroscopy, and

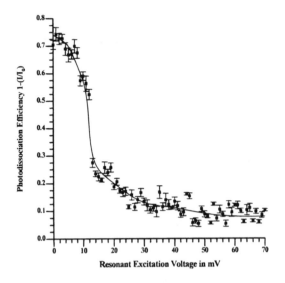

FIGURE 29
Effect of resonant excitation voltage on the axial excursions of protonated diglyme. As the axial excursions of the ions increase, the amount of time the ion packet spends in the radial plane of the ring electrode diminishes significantly. (Reproduced with permission of Elsevier Science Publishers from *J. Am. Soc. Mass Spectrom.*, vol. 5, 1994.)

photoreactivity. With the development of TIP spectroscopy, the electronic spectra of mass-selected molecular ion clusters, the study of stable forms of complex ions, and investigations into the potential energy surfaces of complex ions can give a detailed look at the fundamental behavior of ions in the gas phase.[29] Although much remains to be done in these areas of research, several new fields which fall under the description of photodissociation have emerged. Perhaps the two most promising areas are ion tomography and the photodissociation of large biomolecules.

Ion tomography provides a basis for experimental investigation into ion cloud shape and ion trajectories in the ion trap. It also serves as experimental verification for theoretical simulation programs, and provides a unique alternative to spectroscopic investigations of ion behavior, where the experimental modification of both the ring electrode and end-caps can create large trapping field imperfections. The DC pulse/laser probe technique could in principle be used for stability diagram mapping of the $q_{z,r}$, $\beta_{z,r}$ relationships. This is particularly important for determining the exact shape of the stability diagram for applications such as apex isolation or the study of nonlinear field effects.[59] The combination of spatially resolved photodissociation, as developed by Hemberger and Williams,[38] and the DC pump/laser probe experiments of Lammert et al.[43] could provide a very powerful technique for the investigation of ion trajectories in the quadrupole ion trap.

The most important development in photodissociation ion trap research could come from the photodissociation of large biomolecules. Typically, CID for larger-mass ions (thousands of daltons) becomes increasingly difficult due to several factors. The two most important are the inefficiency of energy transfer observed between the low-mass collision gas and the high-mass analyte, and the inability to impart enough energy to the ion of interest without causing ejection or scattering. Provided that the cross-section for photodissociation is large enough, high-energy laser irradiation or a multiphoton absorption process, combined with the long storage times associated with the ion trap, has the potential for fragmenting ions in the several thousand-dalton range and beyond.

The use of photodissociation in ICR using high-energy photons (e.g., 193 nm, 6.4 eV) was demonstrated successfully by Williams et al.[60] Photodissociation was found to have the added advantage of photofragment production in the center of the ICR cell with little excess kinetic energy deposition, thus yielding high photodissociation efficiencies. Consecutive high-energy laser pulses (e.g., 193 nm, 6.4 eV) were used by Bowers et al.[61] to elucidate the structure of oligopeptides in an ICR. As these techniques are developed for the quadrupole ion trap, the ability to increase the photodissociation cross-section with multipass ring electrode designs should help to increase the dissociation efficiencies of large biomolecules.

REFERENCES

1. Busch, K.L.; Glish, G.L.; McLuckey, S.A. in *Mass Spectrometry/Mass Spectrometry: Techniques and Applications of Tandem Mass Spectrometry*; VCH: New York, 1988; pp. 87–90.
2. Dunbar, R.C. in *Gas Phase Ion Chemistry, Vol. 2*; M.T. Bowers, Ed.; Academic Press: London, 1979; pp. 182–220.
3. Dunbar, R.C. in *Lecture Notes in Chemistry*; H. Hartmann and K.P. Wanczeck, Eds.; Springer-Verlag: Berlin, 1982; pp. 1–26.
4. Dunbar, R.C., in *Gas Phase Ion Chemistry, Vol. 3: Ions and Light*; M.T. Bowers, Ed.; Academic Press: London, 1984; pp. 129–166.
5. Rosenfeld, R.N.; Jasinski, J.M.; Brauman, J.I. *J. Am. Chem. Soc.* 1979, 101, 3999–4000.
6. Bomse, D.S.; Woodin, R.L.; Beauchamp, J.L. *J. Am. Chem. Soc.* 1979, 101, 5503–5512.
7. Thorne, L.R.; Beauchamp, J.L. in *Gas Phase Ion Chemistry. Vol. 3: Ions and Light*; M.T. Bowers, Ed., Academic Press: London, 1984; pp. 41–97.
8. Tumas, W.; Foster, R.F.; Brauman, J.I. *Isr. J. Chem.* 1984, 24, 223–231.
9. Drzaic, P.S.; Marks, J.; Brauman, J.I. in *Gas Phase Ion Chemistry, Vol. 3: Ions and Light*; M.T. Bowers, Ed., Academic Press: London, 1984; pp. 167–211.
10. Mead, R.D.; Lineberger, W.C. in *Gas Phase Ion Chemistry, Vol. 3: Ions and Light*; M.T. Bowers, Ed., Academic Press: London, 1984; pp. 213–248.
11. van der Hart, W.J. *Mass Spectrom. Rev.* 1989, 8, 237–268.
12. van der Hart, W.J. *Int. J. Mass Spectrom. Ion Processes.* 1992, 118/119, 617–633.
13. Hughes, R.J.; March, R.E.; Young, A.B. *Int. J. Mass Spectrom. Ion Phys.* 1982, 42, 255–263.
14. Hughes, R.J.; March, R.E.; Young, A.B. *Can. J. Chem.* 1983, 61, 824–833.
15. Hughes, R.J.; March, R.E.; Young, A.B. *Can. J. Chem.* 1983, 61, 834–845.
16. March, R.E.; Hughes, R.J. *Quadrupole Storage Mass Spectrometry*; John Wiley & Sons: New York, 1989, and references within.
17. March, R.E.; Kamar, A.; Young, A.B. in *Advances in Mass Spectrometry 1985: Proc. 10th Int. Mass Spectrom. Conf.*; Wales, 1986, pp. 949–950.

18. Kamar, A. Application of the Quadrupole Ion Storage Trap (QUISTOR) to the Study of Gas Phase Ion/Molecule Reactions, Ph.D. thesis, Queen's University, Kingston, Ontario, Canada, 1985.

19. March, R.E.; Young, A.B.; Hughes, R.J.; Kamar, A.; Baril, M. *Spectrosc. Int. J.* 1984, 3, 17–32.

20. Kamar, A.; Young, A.B.; March, R.E. *Can. J. Chem.* 1986, 64, 1979–1988.

21. Young, A.B.; March, R.E.; Hughes, R.J. *Can. J. Chem.* 1985, 63, 2324–2331.

22. March, R.E.; Hughes, R.J.; Young, A.B. in *Proc. 13th Meet. British Mass Spectrometry Society*, Warwick, U.K., 1983, pp. 77–79.

23. Louris, J.N.; Brodbelt, J.S.; Cooks, R.G. *Int. J. Mass Spectrom. Ion Processes.* 1987, 75, 345–352.

24. Louris, J.N.; Cooks, R.G.; Syka, J.E.P.; Kelley, P.E.; Stafford, G.C.; Todd, J.F.J. *Anal. Chem.* 1987, 59, 1677–1685.

25. Creaser, C.S.; McCoustra, M.R.S.; O'Neill, K.E. *Org. Mass Spectrom.* 1991, 26, 335–338.

26. Uechi, G.T.; Dunbar, R.C. *J. Chem. Phys.* 1990, 93, 1626–1631.

27. Mikami, N.; Miyata, Y.; Sato, S.; Toshiki, S. *Chem. Phys. Lett.* 1990, 166, 470–474.

28. Mikami, N.; Sato, S.; Ishigaki, M. *Chem. Phys. Lett.* 1991, 180, 431–435.

29. Mikami, N.; Sato, S.; Ishigaki, M. *Chem. Phys. Lett.* 1993, 202, 431–436.

30. Mather, R.E.; Waldren, R.M.; Todd, J.F.J.; March, R.E. *Int. J. Mass Spectrom. Ion Phys.* 1980, 33, 201–206.

31. Benilan, M.-N.; Audoin, C. *Int. J. Mass Spectrom. Ion Phys.* 1973, 11, 421–425.

32. Rosenstock, H.M.; Draxl, K.; Steiner, B.W.; Herron, J.T. *J. Chem. Phys. Ref. Data.* 1977, 6 Suppl., 1.

33. Tripathi, G.N.R.; Schuler, R.H. *Chem. Phys. Lett.* 1982, 88, 253–255.

34. Tripathi, G.N.R.; Schuler, R.H. *J. Chem. Phys.* 1984, 81, 113–121.

35. Kimura, K.; Katsmata, S.; Achiba, Y.; Yamazaki, T.; Iwata, S. *Handbook of HeI Photoelectron Spectra of Fundamental Organic Molecules*, Japan Scientific Society Press: Tokyo, 1981.

36. Ripoche, X.; Dimicoli, I.; LeCalve, J.; Piuzzi, F.; Botter, R. *J. Chem. Phys.* 1988, 124, 305–313.

37. Walter, K.; Boesel, U.; Schlag, E. *Chem. Phys. Lett.* 1989, 162, 261–268.

38. Hemberger, P.H.; Nogar, N.S.; Williams, J.D.; Cooks, R.G.; Syka, J.E.P. *Chem. Phys. Lett.* 1992, 191, 405–410.

39. Williams, J.D.; Cooks, R.G., Syka, J.E.P.; Hemberger, P.H.; Nogar, N.S. *J. Am. Soc. Mass Spectrom.* 1993, 4, 792–797.

40. Stafford, G.C.; Kelley, P.E.; Syka, J.E.P.; Todd, J.F.J. *Int. J. Mass Spectrom. Ion Processes.* 1984, 60, 85–89.

41. Whetten, N.R. *J. Vac. Sci. Technol.* 1974, 11, 515.

42. Kreyszig, E. *Advanced Engineering Mathematics*, 6th ed.; Wiley: New York, 1988; p. 129.

43. Lammert, S.A.; Cleven, C.D.; Cooks, R.G. *J. Am. Soc. Mass Spectrom.* 1994, 5, 29–36.

44. Lammert, S.A.; Cooks, R.G. *J. Am. Soc. Mass Spectrom.* 1991, 2, 487–491.

45. Wang, Y.; Franzen, J. *Int. J. Mass Spectrom. Ion Processes.* in press.

46. Reiser, H.-P.; Julian, R.K.; Cooks, R.G. *Int. J. Mass Spectrom. Ion Processes.* 1992, 121, 49–63.

47. Stephenson, J.L. Jr.; Booth, M.M.; Shalosky, J.A.; Eyler, J.R.; Yost, R.A. *Proc. 41st ASMS Conf. Mass Spectrometry and Allied Topics*, San Francisco, CA, May 30 to June 4, 1993.

48. Stephenson, J.L. Jr.; Booth, M.M.; Shalosky, J.A.; Eyler, J.R.; Yost, R.A. *J. Am. Soc. Mass Spectrom.* in press.

49. White, J.U. *J. Opt. Soc. Am.* 1942, 32, 285–288.

50. Watson, C.H.; Zimmerman, J.A.; Bruce, J.E.; Eyler, J.R. *J. Phys. Chem.* 1991, 95, 6081–6086.

51. Peiris, D.M.; Cheeseman, M.A.; Ramanathan, R.; Eyler, J.R. *J. Phys. Chem.* 1993, 97, 7839–7843.

52. Woodin, R.L.; Bomse, D.S.; Beachamp, J.L. *Chem. Phys. Lett.* 1979, 63, 630–636.
53. Jasinski, J.M.; Rosenfeld, R.N.; Meyer, F.K.; Brauman, J.I. *J. Am. Chem. Soc.* 1982, 104, 652–658.
54. Rosenfeld, R.N.; Jasinski, J.M.; Brauman, J.I. *J. Am. Chem. Soc.* 1982, 104, 658–663.
55. Watson, C.H. Infrared Multiphoton Dissociation of Gaseous Ions Studied by Fourier Transform Ion Cyclotron Resonance Mass Spectrometry, Ph.D. thesis, University of Florida, Gainesville, 1986.
56. Ibuki, T.; Sugita, N.J. *Chem. Phys.* 1983, 79, 5392–5395.
57. Dobler, W.; Ramler, H.; Villinger, H.; Howorka, F.; Lindinger, W. *Chem. Phys. Lett.* 1983, 97, 553–556.
58. Ferguson, E.E.; Adams, N.G.; Smith, D.; Alge, E.J. *J. Chem. Phys.* 1984, 80, 6095–6098.
59. Eades, D.M.; Yost, R.A. *Rapid Commun. Mass Spectrom.* 1992, 6, 573.
60. Williams, E.R.; Furlong, J.J.P.; McLafferty, F.W. *J. Am. Soc. Mass Spectrom.* 1990, 1, 288–294.
61. Bowers, W.D.; Delbert, S.S.; McIver, R.T., Jr. *Anal. Chem.* 1986, 58, 969–972.

Chapter 6

LASER DESORPTION IN A QUADRUPOLE ION TRAP

Jennifer S. Brodbelt, Rafael R. Vargas, and Richard A. Yost

CONTENTS

Part A. Laser Desorption Ion Trap Mass Spectrometry,
by Jennifer S. Brodbelt 206

  I. Introduction 206

 II. Fiberoptic Interface 207

III. Optimization of Laser Desorption 209

 IV. Desorption and Reactions of Metal Ions 210

  V. Desorption of Halide Ions 212

 VI. Desorption of Biological Molecules 212

VII. Laser Desorption with Ion/Molecule Reactions 213

VIII. Conclusions 216

Part B. Laser Desorption and Ion Trap Tandem Mass
Spectrometry, by Rafael R. Vargas and Richard A. Yost ... 217

 IX. Introduction to Laser Desorption and Matrix-Assisted Laser
Desorption Ionization 217
      A.   Overview 217
      B.   Advantages of Laser Desorption Ionization/Ion Trap
           Mass Spectrometry 217

0-8493-4452-2/95/$0.00+$.50

X.   Laser Desorption in Ion Traps 218
    A.   Characterization of Laser Desorption Ionization in
        Ion Traps 218
        1.   Overview of Instrumental Setup 218
        2.   Effect of Buffer Gas Pressure on Trapping
              Efficiency of Laser Desorbed Ions 220
        3.   Effect of Radiofrequency Phase Angle on
              Trapping Efficiency of Laser Desorbed Ions 221
        4.   Effect of Low Mass Cutoff on Trapping
              Efficiency of Laser Desorbed Ions 224
    B.   Increasing Sensitivity and Selectivity for Laser
        Desorption in Ion Traps Using Matrix-Assisted Laser
        Desorption Ionization and Tandem Mass
        Spectrometry 225
        1.   Matrix-Assisted Laser Desorption Ionization ... 225
        2.   Sample Preparation for Matrix-Assisted
              Laser Desorption Ionization 225
        3.   Application of Tandem Mass Spectrometry
              in the Ion Trap to Laser Desorbed Ions 226
    C.   Applications of Laser Desorption in Ion Traps 228
        1.   Biomolecules and the Appearance of
              Adduct Ions 228
        2.   Determining Spiperone in Matrigel 230

XI.   Conclusions and Future of Laser Desorption in Ion Traps ... 231

References 232

PART A
LASER DESORPTION ION TRAP MASS SPECTROMETRY

Jennifer S. Brodbelt

I. INTRODUCTION

Laser desorption/quadrupole ion trap mass spectrometry (ITMS) offers great potential for the ultrasensitive analysis of large molecules, and great inroads have been made toward achieving such a goal since 1988. A pulsed infrared (IR) laser technique in which desorbed metal ions were injected into a quadrupole ITMS was reported in 1989.[1] The injection of gold and tantalum ions was demonstrated with this arrangement, which used a three-element einzel lens to transmit ions into the trap, but the technique was not utilized for desorption of biomolecules. In the same year, IR desorption/ionization of organics was performed inside the cavity of a quadrupole ion trap.[2] This method involved directing the beam of a CO_2 laser through a hole drilled in the ring electrode to a sample probe. Cation-attachment products for several small biomolecules including sucrose and leucine-enkephalin were presented; however, the applicability of this technique was limited by space-charge effects and the low mass range of the quadrupole ion trap detector. In another report,[3] it was shown that collisionally activated dissociation (CAD) could be performed on laser-desorbed salt ions in an ITMS by using an IR laser desorption method similar to the one described in Reference 2. More recently it was shown that ultraviolet matrix-assisted laser desorption could be performed directly in the cavity of the quadrupole ion trap[4] by using a design similar to that described in Reference 3, but in conjunction with the frequency-quadrupled output of a Nd:YAG laser.

In this chapter, the design of a pulsed IR laser desorption technique in conjunction with a quadrupole ITMS, that utilizes external ion formation and axial introduction of ions without auxiliary injection optics, is described.[5] The method employs a fiberoptic laser probe interface that alleviates the need for strategically placed optical windows, and often can provide versatility and portability relative to a conventional optical assembly. Laser desorption by means of a fiberoptic interface was previously shown in a sector mass spectrometer[6] and in a triple quadrupole mass spectrometer,[7] and photodissociation in a quadrupole ion trap[8] was accomplished with a fiberoptic probe. More recently a fiberoptic probe demonstrated for laser desorption/ionization with a Fourier transform-ion cyclotron resonance (FT-ICR) mass spectrometer[9] afforded many of the features most important for an interface suitable for a quadrupole ITMS. The laser desorption method is demonstrated to be useful for several types of experiments, including desorption of organic ions and ion/molecule reactions of selected desorbed cations and anions. The effects of several parameters, including buffer gas pressure, ion storage time, and the trapping voltage, on the effectiveness of ion storage and detection are also described.

II. FIBEROPTIC INTERFACE

The laser desorption probe consisted of a rotatable stainless steel sample tip which was mounted via a screw mechanism on a $\frac{3}{8}$-in. stainless steel probe shaft sealed to a $\frac{1}{2}$-in. probe shaft. The $\frac{1}{2}$-in. probe fits the standard solids probe port of a Finnigan MAT Ion Trap Mass Spectrometer (ITMS™) (see Figure 1), and the probe was admitted through a bellows-valve vacuum interlock. A fiberoptic guide tube was used to direct the fiberoptic to point at the sample tip. The sample probe and fiberoptic

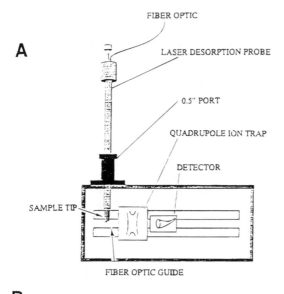

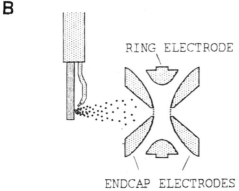

FIGURE 1
Fiberoptic laser desorption interface. Inset: Detail of sample tip orientation.

were positioned orthogonally about 1 cm outside the ion trap. The end-cap electrode had multiple holes for improved transmission of laser-desorbed species. The plume of desorbed material emitted from the surface was aligned directly with the holes in the end-cap electrode. A Nd:YAG laser operated in the Q-switch mode at both 532 and 1064 nm, provided the desorption pulse and power densities were adjusted from 2×10^6 W cm^{-2} to 2×10^8 W cm^{-2}, depending upon the desired type of laser desorption. This arrangement resulted in typical pulse energies of 7 to 20 mJ as measured at the outlet end of the fiberoptic. The laser beam was focused into the fiberoptic by a single lens of focal length 80 mm. The fiberoptic was Teflon clad and had a core diameter of 600 μm.

With each laser shot, ions and neutrals were desorbed from the sample tip, drifted through the end-cap electrode, and some of the ions were trapped. The laser spot size on the surface is about 0.25 mm². The rotatable sample tip allowed >1000 laser shots, each on a fresh surface. At the start of each computer controlled scan, the triggering of the laser was followed by a delay time of 0 to 500 ms in order to allow collisional cooling and/or reactions of ions. Finally, the amplitude of the radiofrequency (RF) voltage applied to the ring electrode was increased so as to eject ions in order of increasing m/z ratio onto an externally located electron multiplier detector. For those compounds with mol wt >650 Da, the resonant ejection mass range extension mode was used.

III. OPTIMIZATION OF LASER DESORPTION

Pulsed IR laser desorption results in formation of ions that typically have an average distribution of kinetic energies from <1 eV to a few electronvolts for salt ions[10] and organic ions, and up to >50 eV for metal ions[10] formed from plasma ignition. The experimental conditions which optimized the trapping of ions generated externally to the ion trap were evaluated. The parameters of particular significance included the helium buffer gas pressure, which is usually maintained at 1 m torr for conventional quadrupole ITMS experiments, the storage time in which ions undergo collisional cooling in the ion trap, the position of the probe tip relative to the transmission holes into the ion trap, and the RF voltage amplitude applied to the ring electrode which creates the quadrupole trapping field.

It was found that the optimal position of the probe relative to the end-cap electrode required alignment of the probe surface with the transmission holes in the end-cap. After each laser pulse, the primary plume of desorbed material was ejected 90° from the sample surface. Thus, this plume was aligned to pass directly through the end-cap electrode. A high buffer gas pressure was found to assist in the trapping of the desorbed species, and typically 1 to 4 m torr helium was introduced, depending on the sample. Presumably, the high helium pressure enhanced collisional damping of the translational energies of the desorbed species such that they were trapped upon admission through the holes in the end-cap electrode. The kinetic energy damping effect, whereby ions underwent collisional cooling during their residence in the ion trap, was influenced by both the helium pressure and the storage time prior to detection. When a high helium pressure was used (2 mtorr), long storage times >50 ms offered only modest advantages for effective ion trapping; however, at the lower helium pressures (≤1 mtorr), it appeared that a delay of 100 to 400 ms afforded a fivefold enhancement of ion detection. Under the latter conditions, ions had time to undergo kinetic cooling to the center of the trap from which they were ejected more effectively and detected during the analytical scan. Depending on the combination of pressure and cooling time used, these conditions corresponded to an ion undergoing from 10^3 to 5×10^3 collisions prior to detection.

The amplitude of the RF voltage applied to the ring electrode during the laser desorption pulse and ion cooling interval (delay time) had a dramatic effect on the storage and detection of ions. When the amplitude was less than about 100 V_{0-p} or >1000 V_{0-p} during the desorption pulse, the trapping of ions was quenched. Apparently, at very low voltages (<50 V_{0-p}) the trapping field strength was insufficient to retain laser-desorbed species. This behavior paralleled the trend observed for operation of the quadrupole ITMS when using conventional ionization modes (electron ionization, EI; chemical ionization, CI) in which ions were formed inside the trap. At RF voltage amplitudes >1000 V_{0-p}, desorbed ions apparently either could not penetrate the potential barrier to the quadrupole field or they became too highly energetic to be retained in the trapping field. Based on these experimental observations, the RF voltage was held at 400 V_{0-p} during the laser desorption event and cooling period, the helium pressure was maintained at >1 mtorr, and a cooling period of 300 ms was used prior to detection.

Despite the very different ranges of kinetic energies for the desorbed ions (alkali metal ions from salts vs. cationized organics vs. metal ions by plasma ignition), it was found that the conditions described above were most suitable for trapping the largest array of ions. This situation was rationalized by assuming that the desorbed ions had a range of kinetic energies, and only those with the appropriate overlap of kinetic energies with the established storage conditions were trapped. For example, the alkali metal ions formed from salts were retained under virtually any combination of trapping conditions, whereas the cationized organics typically required the highest helium pressures and longest delay times for optimal detection. After plasma ignition ionization of metals, only the low kinetic energy tail of transition metal ions was ever effectively trapped, and the majority of hotter ions were too energetic to be retained.

IV. DESORPTION AND REACTIONS OF METAL IONS

At low power densities (6×10^6 W cm^{-2}),[11] alkali metal ions were formed from desorption of any salt applied to the probe tip. These ions were trapped and used for subsequent cation attachment experiments with organic substrates introduced into the vacuum chamber. For example, potassium ions were formed very efficiently by direct laser desorption of KBr salt applied to the probe tip. These potassium cations underwent selective attachment to organic substrates, as shown in Figure 2A for formation of adducts of 12-crown-4 and triethylene glycol dimethylether (designated as M_1 and M_2, respectively). After isolation of the potassium-bound adduct (Figure 2B) by applying an appropriate combination of direct current (DC) and RF voltages, i.e., apex isolation mode, CAD of the $(M_1 + K + M_2)^+$ ion by application of a supplementary resonant voltage resulted in cleavage of the ether/alkali metal electrostatic bonds, forming $(M_1 + K)^+$ and $(M_2 + K)^+$ products (Figure 2C). The abundances of the two prod-

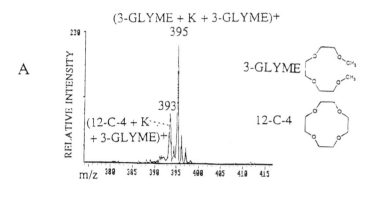

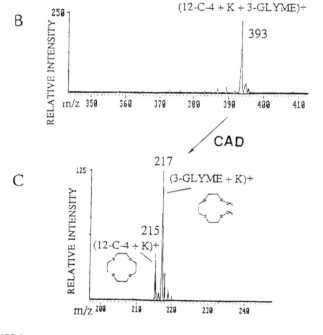

FIGURE 2
(A) Complex formation between 12-crown-4 and triethylene glycol dimethylether; (B) isolation of postassium-bound complex of 12-crown-4 and triglyme; (C) CAD of the adduct shows that triethylene glycol dimethylether has a higher potassium ion affinity than 12-crown-4.

ucts gave an indication of the relative potassium ion affinities of each ether, and the higher abundance of the triethylene glycol dimethylether product showed that it had a greater potassium ion affinity than did 12-crown-4.

At higher power densities (10^8 W cm^{-2}), transition metal ions were desorbed either from the stainless steel probe surface (Figure 3A) or from metal foils attached to the probe tip. These metal species were used to ionize various organic substrates, such as aromatic compounds, by simple attachment and/or insertion processes. An example of desorption of metal ions with subsequent attachment to naphthalene to form $(M + Cr)^+$ and $(M + Fe)^+$ ions, and then metal-bound dimers is shown in Figure 3B and C.

V. DESORPTION OF HALIDE IONS

Negative ions are difficult to form in the quadrupole ITMS by conventional methods, such as electron attachment ionization, because there is no ready source of thermal electrons. This difficulty is attributed to the influence of the high RF field on the electron population generated by a filament assembly. However, the laser desorption method afforded a relatively simple way to form negative ions in the quadrupole ITMS. In the negative ion mode, halide anions from the alkali metal halide salt may be desorbed and detected, as shown for Br$^-$ in Figure 4A. These anions underwent attachment reactions with organics, as indicated in Figure 4B for 15-crown-5.

VI. DESORPTION OF BIOLOGICAL MOLECULES

At moderate power densities (10^7 W cm^{-2}), organic ions, often cationized species, were formed and trapped. Several representative spectra are shown in Figures 5 to 8. The organic substrates include digitoxigenin (Figure 5), tetraphenylporphine (Figure 6), leucine-enkephalin (Figure 7), and gramicidin S (Figure 8). With the mass-range extension mode of operation, i.e., with the application of a supplementary RF voltage of 29 $V_{(p-p)}$ at 150 kHz across the end-cap electrodes, the largest molecule that was analyzed was gramicidin S, $(M + K)^+$ at 1180 Da. Typically, cationized molecular ions with little fragmentation were observed in all of the mass spectra, except for the mass spectrum showing the desorption of tetraphenylporphine in which molecular ions and two fragment ions were seen (Figure 6). The fragment ion at m/z 537 was due to loss of one phenyl group from the molecular ion, but the identity of the other fragment observed at nominally m/z 473 was not elucidated. In Figure 7 both protonated and various cationized forms of leucine-enkephalin were observed.

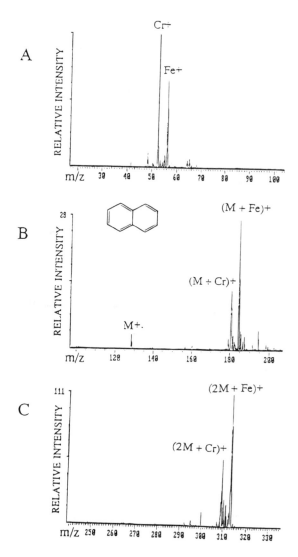

FIGURE 3
(A) Desorption of transition metal ions (Cr⁺ and Fe⁺); (B) metal reactions with naphthalene; (C) formation of metal-bound dimers of naphthalene.

VII. LASER DESORPTION WITH ION/MOLECULE REACTIONS

Laser desorption also provides a means of generating novel reagent ions for selective ion/molecule reactions, such as were described briefly above for halide ion attachments and metal ion reactions. Laser-desorbed ions have already proven useful for various CI methods, including the distinction of isomers.[12] An example of the use of laser-desorbed metal

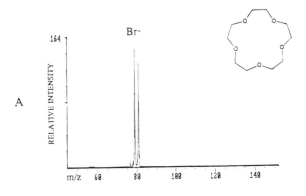

A

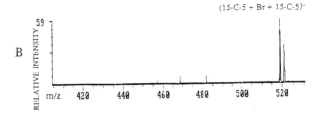

B

FIGURE 4
(A) Desorption of bromine ions from KBr; (B) reactions with 15-crown-5 to form dimers.

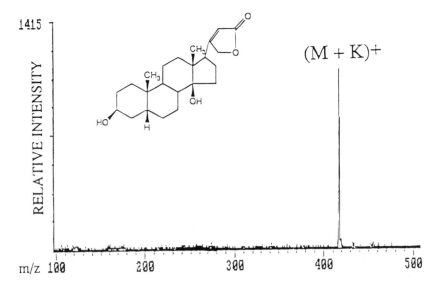

FIGURE 5
Desorption of digitoxigenin (mol wt 374, in a KBr mixture) from the probe.

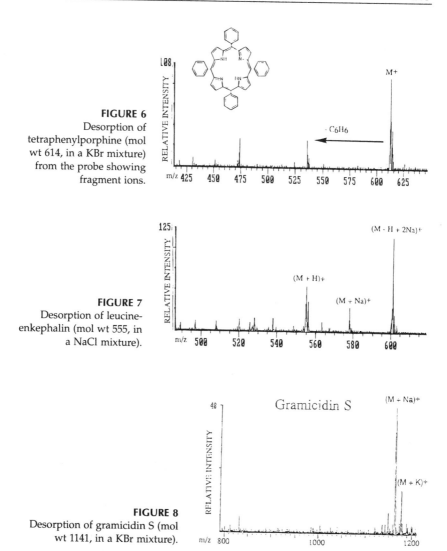

FIGURE 6
Desorption of
tetraphenylporphine (mol
wt 614, in a KBr mixture)
from the probe showing
fragment ions.

FIGURE 7
Desorption of leucine-
enkephalin (mol wt 555, in
a NaCl mixture).

FIGURE 8
Desorption of gramicidin S (mol
wt 1141, in a KBr mixture).

ions for characterization of related molecules is shown in Figure 9. Iron ions were produced by desorption from an iron foil and allowed to interact with different polycyclic aromatic hydrocarbons, including anthracene and pentacene.[13] As illustrated in Figure 9A, the iron ions underwent charge exchange or formed an iron complex with anthracene, but only charge exchange was observed for pentacene (Figure 9B). Pentacene has a much lower ionization energy than anthracene (6.6 vs. 7.5 eV),[14] and thus the charge exchange reaction involving the iron ions is more exothermic (−0.35 eV as compared to −1.25 eV). It was suggested

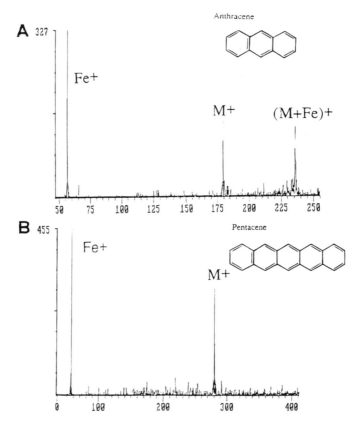

FIGURE 9
Reactions of iron ions with (A) anthracene and (B) pentacene.

that the absence of the iron/pentacene complex reflected that the interaction between the pentacene molecules and iron ions resulted in complexes that were too energized internally to remain intact for a sufficient
period of collisional stabilization. One advantage of using laser-desorbed
reagent ions for ion/molecule reactions is the absence of a continuous
supply of reagent neutrals (as is present in conventional CI techniques)
to engage in competitive charge transfer reactions with analyte ions.

VIII. CONCLUSIONS

The ability to perform laser desorption in conjunction with quadrupole ion trap mass spectrometry has opened up many new avenues of
research, including those in biological mass spectrometry and organometal-

lic chemistry. Moreover, the unparalleled sensitivity of the quadrupole ion trap makes it an especially attractive device for studies involving trace analysis.

PART B
LASER DESORPTION AND ION TRAP TANDEM MASS SPECTROMETRY

Rafael R. Vargas and Richard A. Yost

IX. INTRODUCTION TO LASER DESORPTION AND MATRIX-ASSISTED LASER DESORPTION IONIZATION

A. Overview

Laser desorption ionization (LDI)/quadrupole ITMS is a relatively new method that has received much attention because of its great potential for becoming a highly sensitive and selective method for analysis of large molecules in the solid phase. Research performed over many years using other types of mass spectrometers have shown LDI to be a simple ionization method applicable to many different types of solid samples, including thermally labile samples. Many examples have appeared in the recent literature demonstrating the production of intact molecular ions from small compounds to large biological compounds and polymers. Specifically, the method of matrix-assisted laser desorption ionization (MALDI), as first introduced by Hillenkamp and Karas in 1987, makes possible routine analysis of peptides, proteins, and other biomolecules.[15-18]

B. Advantages of Laser Desorption Ionization/Ion Trap Mass Spectrometry

Quadrupole ITMSs have been used to mass analyze the ions produced by many different ionization methods (for example, electrospray ionization).[19] Ions that are formed within the ion trap, or that are injected into the trap, can be studied using the high sensitivity and selectivity inherent to the ion trap.[20] The quadrupole ITMS can trap and store ions of all masses from a pulsed laser desorption event; in contrast, quadrupole mass spectrometers and other scanning instruments are not able to obtain spectra over a wide mass range for a pulsed ionization event. All ions produced by the laser pulse can be detected with the quadrupole ITMS regardless of the duration of the laser desorption event.

Much of the analysis of very large molecules using laser desorption mass spectrometry has been performed using time-of-flight (TOF) mass spectrometry. Using different methods of dissociation, several researchers have demonstrated MS/MS using dual TOF stages of mass analysis,[21-23] a capability not yet readily available. Tandem mass spectrometry is a powerful and widely accepted method for trace mixture analysis, and it is particularly suitable where chromatographic separation is not possible. Research in our group and others has demonstrated the ability of MS/MS (typically with triple quadrupole mass spectrometers) to identify rapidly trace level components in complex mixtures with little or no sample preparation or chromatographic separation. Such analytical capabilities are particularly attractive for imaging microanalysis, where extraction or chromatographic separation is clearly not applicable.[24] The quadrupole ion trap has the advantage of being able to perform multiple stages of mass analysis (MS^n) without the need for modification of the system or addition of hardware.

The quadrupole ion trap offers several advantages over other types of mass spectrometers for MS/MS. Perhaps most important for the LDI/MS/MS experiment is the ability to trap the ions produced by the laser pulse for MS/MS analysis, a capability shared with other stored ion mass spectrometers such as the FT-ICR mass spectrometer. Less stringent vacuum requirements and smaller instrument size are offered by the quadrupole ion trap in contrast to the FT-ICR instruments. The ion trap is a much simpler and more compact instrument than are sector, hybrid sector/quadrupole, or triple quadrupole MS/MS instruments. Finally, recent research has demonstrated dramatic increases in the mass range[25,26] of the ion trap as well as its mass resolution.[27-29] These advances in ion trap technology have greatly increased the ability of the ion trap to analyze large molecules.[30-32]

X. LASER DESORPTION IN ION TRAPS

A. Characterization of Laser Desorption Ionization in Ion Traps

1. Overview of Instrumental Setup

The layout of the LDI/quadrupole ITMS system used in the work described below is based on a Finnigan MAT ITMS™, as shown in Figure 10. This system is similar to those previously reported by Cotter et al.[2] and Goeringer et al.[3] The beam from a pulsed nitrogen laser (337 nm) is directed through ultraviolet-grade focusing lens and a low distortion quartz window, which is mounted inside a vacuum flange. To enter the ion trap, the laser beam passes through the RF ring electrode via a $\frac{1}{8}$-in. diameter hole and impinges on the sample surface. The sample is introduced into the manifold on the tip of a stainless steel probe through a flange-mounted vacuum probe lock. The tip of the sample probe is then introduced into the ion trap through a $\frac{1}{8}$-in. diameter hole opposite the laser beam entry hole. A Teflon tip holder provides electrical insulation between the high RF voltages and the stainless steel probe when the sample surface is placed flush with the ring electrode surface. The two $\frac{1}{8}$-in. holes that were machined into the RF ring electrode did not cause any noticeable change in the ion trapping efficiency or mass resolution of the ion trap.

The laser is triggered by the ITMS electronics so that a single laser pulse occurs during the trap-and-store time of each scan. The typical scan duration used in the ITMS is 100 ms. A scan duration of this magnitude corresponds well to the repetition rate of the nitrogen laser (maximum 20 Hz), which gives a laser energy of 250 µJ per pulse. The LDI scan sequence is shown in Figure 11. Other parameters involved in operating the ion trap for LDI are similar to those typically employed for EIMS.

A base pressure of 8×10^{-8} torr is maintained inside the manifold by a turbomolecular pump. The pressure is increased to 1 mtorr (corrected) by introducing helium gas into the manifold through a leak valve. The helium acts as a buffer to help trap and collisionally cool ions in the trap. The ions trapped during the laser pulse can then be ejected from the trap

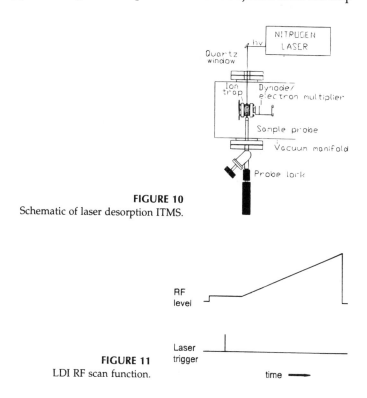

FIGURE 10
Schematic of laser desorption ITMS.

FIGURE 11
LDI RF scan function.

for detection using the mass-selective instability method.[33] During the period when the RF voltage amplitude is being ramped, ion detection occurs with an electron multiplier located outside the ion trap. The detector signal is then sent to the ITMS electronics and recorded using a personal computer.

2. Effect of Buffer Gas Pressure on Trapping Efficiency of Laser Desorbed Ions

The buffer gas pressure was found to have a significant effect on ion trapping efficiency, as seen in Figure 12A for the isotopes of copper.[34] In this experiment, copper metal used as the sample probe tip material was ionized by LDI in order to produce a consistent ion signal over hundreds of laser pulses. The trapped ion intensities of the copper isotopes increased with increasing helium buffer gas pressure. This behavior is similar to that seen in EI in the ion trap, except that at very low pressures few if any ions are detected by LDI. This observation demonstrates that buffer gas plays a very important role in the trapping of laser desorbed ions produced in the area of the ring electrode surface. Two other buffer gases, nitrogen and argon, were also studied; both were found to affect the ion trapping efficiency as did helium, but at lower pressure ranges.

As seen in Figure 12A, the copper isotope ions increase in intensity at different rates with increasing helium pressure. Therefore, the $^{63}Cu^+/^{65}Cu^+$ isotope ratio varies over the helium buffer gas pressure range, as shown in Figure 12B. The isotope ratio value is also always lower than the natural abundance ratio of 2.3. As was determined in later studies, this observation was due to the fact that the copper metal ions are easily produced in large abundances during the laser pulse event. These large numbers of ions created a space-charge effect which was observed in the peak shapes of the ions. This space-charge effect caused the isotope ratio to be lower than the natural ratio by attenuating the intensity of the lower m/z ($^{63}Cu^+$) ion. After reducing the total number of copper ions formed by reducing the laser irradiance in later experiments, the correct isotope abundance ratio was observed. In a study of various metal ions desorbed from a stainless steel probe tip, the observed isotope ratios for iron, chromium, nickel, and molybdenum were found to be in agreement with literature values when ion production was diminished and space-charge effects were avoided. The buffer gas effects seen here are also important in measuring ion intensity ratios of organic and biological molecules; however, space charging by large ion production can be controlled by reducing the laser irradiance or selectively storing only the ions of interest.

The buffer gas pressure effect on ion intensity is also similar to that observed for molecular ions, and agrees with previously reported data.[34] The important role that buffer gas plays in trapping laser desorbed ions

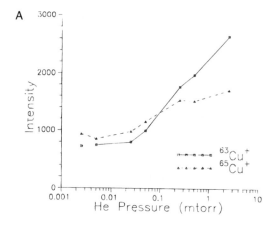

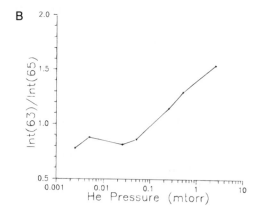

FIGURE 12

(A) $^{63}Cu^+$ and $^{65}Cu^+$ ion intensities vs. helium buffer gas pressure. (B) $^{63}Cu^+/^{65}Cu^+$ isotope ratio vs. helium buffer gas pressure.

(produced at the ring electrode) is analogous to its role for trapping externally produced ions (typically introduced through an end-cap electrode).[1]

After ion trapping has occurred, buffer gas continues to play an important role during ion storage by cooling the ions toward the center of the trap. This process is used to achieve better resolution and sensitivity, as was first discussed by Stafford et al.[33]

3. Effect of Radiofrequency Phase Angle on Trapping Efficiency of Laser Desorbed Ions

The effect of the phase angle of the RF voltage on the trapping efficiency of laser desorbed ions was investigated.[35] In preliminary studies in our laboratory and in studies reported by others, the laser trigger was not synchronized with the RF phase. It was observed that the ion signal varied greatly from pulse to pulse. Due to the short laser desorption event (the laser pulse width is 3 ns in our system compared to the 909 ns period of the 1.1 MHz RF), the ion trapping efficiency might be expected to be dependent on the RF phase. To evaluate this effect, a circuit was built to trigger the laser in synchronization with the RF phase. The circuit also enabled the triggering of the laser at different RF phase angles. The laser pulse was detected by a photodiode and the signal from this photodiode was compared to the RF signal using a digital oscilloscope. The synchronous triggering circuit could vary the delay time between the RF trigger and the laser trigger from 200 to 1200 ns, thus enabling the study of laser pulse triggering over an entire RF period. In order to measure the ion trapping efficiency with respect to the RF phase, a constant source of sample ions is needed over a series of laser pulse events. Toward this end, a graphite rod was used as the sample and probe tip. A mass spectrum of graphite from an average of the first set of ten laser pulses used in the RF phase angle data discussed below is given in Figure 13.

For the RF phase angle study, ten laser desorption mass spectra were taken at each of 29 different points over 1.5 RF cycles, each point accounting for an increment of 50 ns.[35] Each set of ten spectra was averaged to obtain a plot of ion intensity with respect to RF phase angle as seen in Figure 14A. This plot shows the most abundant ion, C_3^+, following the pattern of the RF voltage sine wave (shown in Figure 14B). The ion current pattern is similar to the sine wave except that it has a slower rise and a quicker fall and is slightly shifted in phase.

The C_3^+ ion is much greater in intensity than the other ions detected upon laser irradiation of graphite (as seen in the mass spectrum in Figure 13). However, it was possible in these trials to observe a pattern in the C_4^+ and C_5^+ ion signal patterns which follow a shape similar to that of

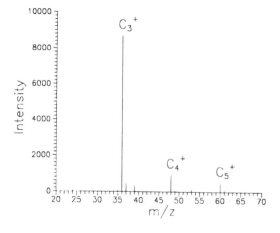

FIGURE 13
LDI mass spectrum of graphite (average of 10 scans).

the C_3^+ ion, although each is shifted in phase angle by different amounts, as shown in Figure 14(C,D). The patterns of these ions are also more sporadic due to their lower signal/noise ratio. A comparison of the ions at different masses raises the question of whether this RF phase dependency of the ion trapping efficiency is also mass dependent, thus complicating the matter of determining the optimum point of laser desorption in the RF phase of the ion trap for any given sample. The next step in determining the degree of mass dependency is to obtain another sample which, like graphite, provides an abundance of different mass ions for multiple laser shots.

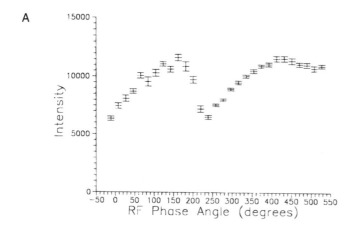

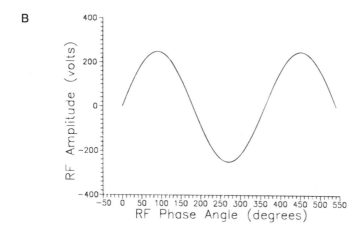

FIGURE 14
(A) C_3^+ ion intensity vs. RF phase angle. (B) Trapping RF amplitude vs. RF phase angle. (C) C_4^+ ion intensity vs. RF phase angle. (D) C_5^+ ion intensity vs. RF phase angle.

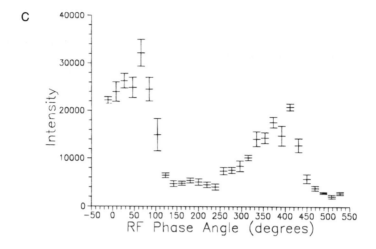

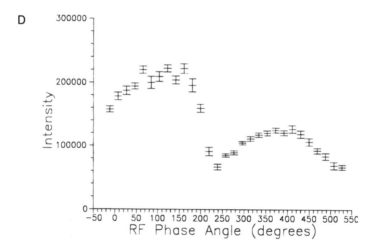

FIGURE 14   CONTINUED

4. Effect of Low Mass Cutoff on Trapping Efficiency of Laser Desorbed Ions

Ions stored in the ion trap are subject to a low mass cut-off; ions of m/z lower than this value have unstable trajectories, while ions of higher m/z will have stable trajectories in the trap. The low mass cutoff level is proportional to the RF amplitude applied to the ring electrode. The effects of this parameter on ion trapping efficiency during trapping and storage periods were studied.[34] The ion intensity of $^{56}Fe^+$ vs. the low mass cutoff level during trapping and storage is shown in Figure 15. As seen

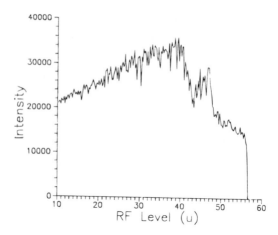

FIGURE 15
$^{56}Fe^+$ ion intensity vs. RF level.

in the plot, the ion intensity varies with the low mass cutoff level, with a maximum occurring at approximately 38 u and complete loss of ion storage above 56 u. Also seen in the plot is a minimum which occurs at approximately 43 u. This loss of trapping efficiency, termed a black hole, arises from octopolar distortions to the quadrupolar trapping field at a stability coordinate ($a_z = 0$, $q_z = 0.69$) corresponding to a value for $\beta_z = \frac{1}{2}$; the black hole has been seen in chemical ionization and in CID experiments, and is currently the focus of some research in the authors' laboratory.[36-38] The overall pattern for ion intensity has been determined to be similar for ion masses <120 u. The effect was found to be different for laser desorbed ions of larger masses such as that of protonated spiperone at 396 u. Evidence suggests that this effect, however, is not a storage effect but a trapping effect. The trapping (or "acceptance") volume at high RF amplitudes is very small for ions that are injected,[39] such as the ions formed at the ring surface. At RF cutoff levels >150 u, ion trapping efficiency for protonated spiperone (m/z 396) does not increase as expected but instead becomes very small. In fact, very few ions are detected when RF cutoff levels are above 150 to 200 u. Although laser-desorbed ions of m/z up to several thousands have been detected in the ion trap, the RF cutoff levels were set low and the ions were detected using axial modulation for mass range extension.[40]

Clearly, there are multiple variables involved in ion trapping including buffer gas pressure, RF phase angle, and low mass cutoff (RF amplitude). Ions desorbed off the sample surface also have a range of velocities (kinetic energies) which "inject" them into the ion trap volume. The combination of these variables ultimately determines the number of desorbed ions that become trapped.

B. Increasing Sensitivity and Selectivity for Laser Desorption in Ion Traps Using Matrix-Assisted Laser Desorption Ionization and Tandem Mass Spectrometry

1. Matrix-Assisted Laser Desorption Ionization

Matrix-assisted laser desorption ionization, as first introduced by Hillenkamp and Karas in 1987, is used to vaporize and ionize involatile and thermally labile organic molecules.[15-18] Although MALDI has been primarily used to study biological molecules with mol wt of 1000 u or greater, in our laboratory the drug compound of interest, spiperone (mol wt 395), was found to be a compound which fragmented easily upon laser ionization and, therefore, was a prime candidate for a softer ionization method.[35] In order to avoid fragmentation, a method of energy transfer must be used which is fast enough to avoid thermal degradation. The fast laser pulses currently used (nanoseconds and shorter) fulfill this requirement. Also, the matrix compound provides efficient absorption of the laser pulse energy, separation of analyte molecules to prevent clustering, and a source of hydrogen for protonation reactions. In effect, the matrix absorbs the laser energy which volatilizes the analyte molecules, and it performs a soft CI of analyte in the selvedge region just above the sample surface. Along with the research performed in recent years to understand better how the method works, MALDI has become quite popular for the study of large biological molecules, as well as many other substances which may require a soft ionization process (synthetic polymers, for example).

2. Sample Preparation for Matrix-Assisted Laser Desorption Ionization

The samples studied in our laboratory were prepared using the following methods. For LDI, 2 to 5 µL of sample solution (0.1 to 1 µg/µL in methanol) were deposited on the probe tip and dried in air before placing in vacuum. For MALDI, 2 to 4 µL of analyte solution (0.1 to 0.5 mM) were deposited on the probe tip followed by an equal volume of the matrix solution (100 to 500 mM) and air dried. Typically, the UV-absorbing matrix compound (sinapinic acid, nicotinic acid, or 2,5-dihydroxybenzoic acid) was dissolved in methanol. Trifluoroacetic acid was added to the matrix solution in a 0.1% concentration in order to promote dissolution of the matrix compound. Through all samples and various concentrations, the analyte-to-matrix ratio was maintained around 1:1000. The proper combination of the analyte and matrix solutions on the probe tip would fulfill the MALDI requirement of a dry, homogeneous mixture of analyte and matrix compound in which analyte molecules are individually surrounded by matrix molecules.[18]

Figures 16 A and B show the LDI and MALDI mass spectra for spiperone (mol wt 395), respectively. The ions that result from LDI without the

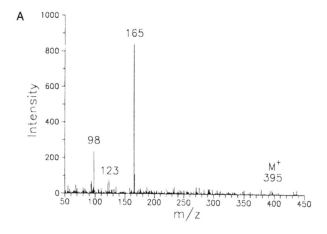

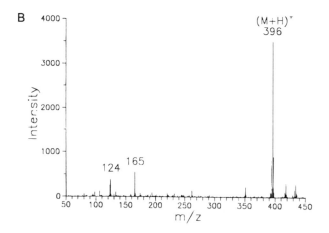

FIGURE 16

(A) LDI mass spectrum of spiperone (mol wt 395). (B) MALDI mass spectrum of spiperone using nicotinic acid matrix (0.5% trifluoroacetic acid) in methanol.

addition of a UV-absorbing matrix are predominantly fragment ions, including m/z 165, 123, and 98. The molecular ion, at m/z 395, is barely discernible. In Figure 16B the protonated spiperone ion at m/z 396 is the predominant species formed in the MALDI experiment using a nicotinic acid matrix. The fragment ion at m/z 165 is also seen. The small peak at m/z 124 is probably protonated nicotinic acid, which is in much greater concentration than spiperone on the sample surface. Some small adduct peaks corresponding to the addition of sodium and potassium ions to the neutral spiperone molecule are also seen in the MALDI spectrum.

An important factor in performing good MALDI is the laser irradiance; published reports indicate that the best results have been obtained when the laser irradiance is around 10^6 W cm^{-2}. Direct measurement of the irradiance of the 3 ns pulse of the nitrogen laser is difficult and would require expensive measurement equipment. Therefore, the laser irradiance was optimized using neutral density filters; the optimum setting for MALDI was determined as the irradiance at which minimal fragment ions were detected and the most abundant peak was the protonated analyte. The laser irradiance could be approximated from given specifications of the laser and the measured spot size. The laser pulse energy is 250 µJ and the pulse period is 3 ns. These values give a pulse peak power of 83 kW. This power, along with the measured spot area of 0.13 mm^2, gives an irradiance of 6.4×10^7 W cm^{-2}. The addition of neutral density filters to obtain the best MALDI mass spectra set the laser beam transmission to <10% of this value, or 6×10^6 W cm^{-2}, fairly close to the previously reported irradiance values. A quick optimization of laser irradiance usually needs to be performed for each new sample prior to obtaining the best data for molecular ion detection.

3. Application of Tandem Mass Spectrometry in the Ion Trap to Laser Desorbed Ions

The additional adduct or matrix ions in the mass spectrum pose less of a problem in the sample analysis with the ion trap than with other mass analyzers because multiple stages of mass analysis can be employed for increased selectivity. Figure 17 shows the daughter mass spectrum obtained by collision-induced dissociation (CID) of protonated spiperone, m/z 396, produced with MALDI. This daughter spectrum is similar to that obtained using other methods of sample introduction and mass analysis, including solids-probe CI in the ion trap and with a triple stage quadrupole instrument.[34]

It is possible to perform MS/MS on the spiperone fragment ions produced in the LDI process by isolating a fragment ion species (m/z 165, for example) and performing CID.[41] As shown in Figure 18, the ions produced from this MS/MS experiment are similar to the fragment ions of spiperone obtained during LDI. The results obtained for MS/MS of laser desorbed ions are similar to those obtained by Goeringer et al.[3] for the cation of an organic salt, trimethylphenylammonium chloride.

C. Applications of Laser Desorption in Ion Traps

1. Biomolecules and the Appearance of Adduct Ions

Leucine-enkephalin (mol wt 555), a five amino acid peptide, was analyzed with the ITMS after ionization using MALDI. Figure 19A shows the MALDI mass spectrum of leucine-enkephalin with a single peak at *m/z* 556 for the protonated molecule. In Figure 19B, the MALDI mass spectrum of leucine-enkephalin is also given, using a different sampling spot than for the previous figure. In this spectrum, we also observed molecular adduct peaks at *m/z* 578 and 600, corresponding to the sodium ion and the carboxyl ion additions to the neutral peptide, respectively. In fact, these adduct ions are in greater abundance than the [M + H]⁺ ion. The sodium probably comes from the sample of leucine-enkephalin. The carboxylic group (from the addition of 45 u), comes from the 2,5-dihydrox-

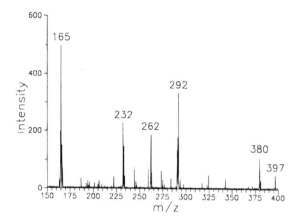

FIGURE 17
MALDI/MS/MS daughter spectrum of protonated spiperone, *m/z* 396 (average of 10 scans).

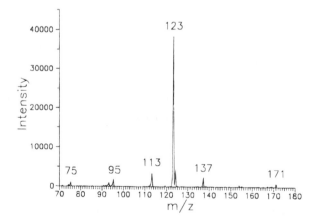

FIGURE 18
MALDI/MS/MS daughter spectrum of *m/z* 165 fragment ion of spiperone (average of 10 scans).

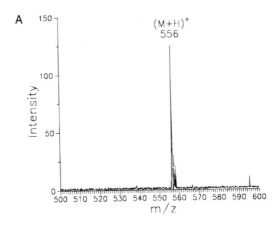

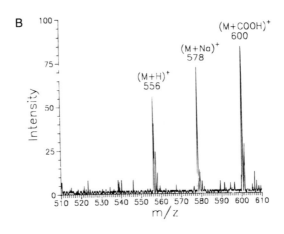

FIGURE 19
(A) MALDI mass spectrum of leucine-enkephalin. (B) MALDI mass spectrum of leucine-enkephalin and adduct ions.

ybenzoic acid matrix compound which is present on the sample surface in a much greater amount (1000 times) than is the peptide.[41]

2. *Determining Spiperone in Matrigel*

For the Matrigel experiments, 0.5 to 1 mg of spiperone (Sigma Chemicals) was mixed in 1 mL of Matrigel (Collaborative Research, Inc.) in the liquid state. Matrigel, a mouse extracellular tissue, is being used here as a model biological matrix. Its composition, as described by the manufacturer, is laminin, collagen IV, heparan sulfate proteoglycans, entactin, and nidogen. For analysis of spiperone in Matrigel, 4 µL of the

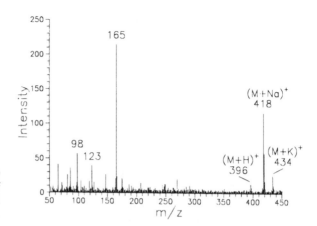

FIGURE 20
MALDI/MS of
spiperone in Matrigel
(average of 20 scans).

mixture was deposited on the sample probe tip followed by 4 μL of matrix solution. The sample does not air dry due to the gelatinous nature of the Matrigel membrane, but does dry upon being placed in high vacuum.

To obtain the mass spectrum shown in Figure 20, the spiperone/Matrigel mixture was deposited on the probe tip and a sinapinic acid matrix solution added. As seen in the figure, there is little production of protonated spiperone; the sodium adduct peak is also approximately ten times greater in abundance. The Matrigel membrane may be the source of the sodium to this ionization process. The relatively abundant spiperone fragment ions indicate that the MALDI process is not optimized. The presence of the Matrigel does not permit the isolation of spiperone by the matrix compound alone; instead, the spiperone molecules are surrounded by both matrix molecules and the Matrigel components. This situation is suggested by the spectrum where both molecular ions and fragment ions of spiperone are seen. The irradiance could not be optimized by reducing it further to eliminate fragmentation because ions were not detected at lower settings. This observation suggests that the laser energy is not as efficiently absorbed as in the experiment using pure spiperone with the matrix compound.[41]

The addition of Matrigel to the sample composition also affects the quality of the MS/MS spectra. In the MS/MS spectra of spiperone in Matrigel, the spiperone ion is incompletely isolated. Smaller peaks appear throughout the mass range of the scan due to ion/molecule reactions between the spiperone ions and late-desorbing neutral species off the sample surface. Increasing the period of the isolation steps until the delayed desorption of neutrals has ceased would remove these additional ions; however, the loss of molecular spiperone ions to these side reac-

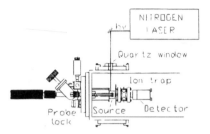

FIGURE 21

Schematic of laser desorption/ion injection/ion trap mass spectrometer.

tions would still reduce the sensitivity of the method for the analyte. Current research is focused on eliminating these effects, including the use of an external LDI source with injection of the ions (but not desorbed neutrals) into the trap for analysis.

If we use the major fragment of spiperone at m/z 165 for analysis, we can determine the structure of this ion with MS/MS, and we can compare these data with those obtained previously and shown in Figure 18. The 165$^+$ fragment ion is much more easily detected than is the molecular species and could be used for detection of spiperone. This, method of analysis is currently being evaluated in the authors' and several other laboratories.

XI. CONCLUSIONS AND FUTURE OF LASER DESORPTION IN ION TRAPS

From the research being performed in a number of laboratories, including ours, it is readily concluded that the ITMS™, combined with lasers, has a promising future. The process of LDI continues to improve in efficiency and ease due to the use of better lasers, new matrices for MALDI, and supplemental ionization.[24,42] Similarly, dramatic new capabilities are being developed for the quadrupole ion trap. The advances on both these fronts promise new opportunities for biomolecular and high mass analyses.

In our laboratory we are addressing the problem of increasing sensitivity by increasing analyte ion production from the sample. LDI and MALDI will be optimized for drug compounds in tissue. The results obtained thus far from the internal laser desorption configuration are promising; we are currently performing studies using an external LDI system. The application of this external source, from which the ions would be introduced to the trap through an ion injection lens system, would eliminate many of the problems encountered with *in situ* ion formation. An external ion source would eliminate the presence of late-desorbing neutrals during the MS/MS process. It would permit optimization of the conditions for CI of laser desorbed analyte neutrals.[24] It would permit selective trapping of the ions of interest with the removal of the matrix ions. Finally, an external source would make laser microscopy easier to achieve with a sampling stage.

The external source configuration has already been developed in our laboratory. As shown in Figure 21, it uses an on-axis ion injection system with the analyzer hardware and electronics of a Finnigan ITS40™. This system has been configured inside a home-built glass-top cradle vacuum manifold. This manifold allows greater accessibility and versatility when developing the external source system than does the conventional ITMS vacuum manifold. The manifold also provides differential pumping between the source and the analyzer/detection assemblies, which was a requirement for particle beam experiments performed in our laboratory. The nitrogen laser beam is introduced at 90° to the ion beam axis and impinges on a sample surface at an angle of 45°. The sample probe is introduced through a standard $\frac{1}{2}$-in. probe lock on-axis with the ion injection system. This instrumentation will be used to perform MS^n experiments on laser desorbed ions as well as on ions produced by CI of laser desorbed neutrals.

An external source configuration also allows the addition of a fine-focusing system in order to achieve laser microscopy and sample surface imaging. These imaging capabilities, currently available on laser/TOF microprobe instruments, would provide a true molecular microprobe when combined with the analytical capabilities of the quadrupole ion trap.

REFERENCES

1. Louris, J.N.; Amy, J.W.; Ridley, T.Y.; Cooks, R.G. *Int. J. Mass Spectrom. Ion Processes.* 1989, 88, 97.
2. Heller, D.N.; Lys, I.; Cotter, R.J.; Uy, M.O. *Anal. Chem.* 1989, 61, 1083.
3. Glish, G.L.; Goeringer, D.E.; Asano, K.G.; McLuckey, S.A. *Int. J. Mass Spectrom. Ion Processes.* 1989, 94, 15.
4. Goeringer, D.E.; Chambers, D.M.; McLuckey, S.A.; Glish, G.L. Presented at the 4th Annu. ASMS Sanibel Conf. Mass Spectrom., January 1992.
5. McIntosh, A.; Donovan, T.; Brodbelt, J. *Anal. Chem.* 1992, 64, 2079.
6. Cechetti, W.; Polloni, R.; Maccioni, A.N.; Traldi, P. *Org. Mass Spectrom.* 1986, 21, 517.
7. Emary, W.B.; Wood, K.V.; Cooks, R.G. *Anal. Chem.* 1987, 59, 1069.
8. Louris, J.N.; Brodbelt, J.S.; Cooks, R.G. *Int. J. Mass Spectrom. Ion Processes.* 1987, 75, 345.
9. Hogan, J.D.; Beu, S.C.; Laude, D.A.; Majidi, V. *Anal. Chem.* 1991, 63, 1452.
10. Van der Peyi, G.; Van der Zand, W.; Kistemaker, P. *Int. J. Mass Spectrom. Ion Processes.* 1984, 62, 51.
11. Vertes, A.; DeWolf, M.; Juhasz, P.; Gijbels, R. *Anal. Chem.* 1989, 61, 1029.

12. Freiser, B.S. *Talanta*. 1985, 8B, 697.
13. Asgharian, N.; Brodbelt, J.S. unpublished results.
14. Lias, S.G.; Bartmess, J.E.; Liebman, J.F.; Holmes, J.L.; Levin, R.D.; Mallard, W.G. *J. Phys. Chem. Ref. Data*. 1988, 17 (Suppl. 1).
15. Karas, M.; Bachmann, D.; Bahr, U.; Hillenkamp, F. *Int. J. Mass Spectrom. Ion Processes*. 1987, 78, 53.
16. Stahl, B.; Steup, M.; Karas, M.; Hillenkamp, F. *Anal. Chem*. 1991, 63, 1463.
17. Zhao, S.; Somayajula, K.V., Sharkey, A.G.; Hercules, D.M.; Hillenkamp, F.; Karas, M.; Ingendoh, A. *Anal. Chem*. 1991, 63, 450.
18. Hillenkamp, F.; Karas, M.; Beavis, R.C.; Chait, B.T. *Anal. Chem*. 1991, 63, 1193A.
19. March, R.E. *Int. J. Mass Spectrom. Ion Processes*. 1992, 118/119, 71.
20. Johnson, J.V.; Yost, R.A. *Anal. Chem*. 1990, 62, 2162.
21. Schey, K.; Cooks, R.G.; Grix, R.; Wöllnik, H. *Int. J. Mass Spectrom. Ion Processes*. 1987, 77, 49.
22. Cornish, T.J.; Cotter, R.J. *Anal. Chem*. 1993, 65, 1043.
23. Jardine, D.R.; Morgan, J.; Alderdice, D.S.; Derrick, P.J. *Org. Mass Spectrom*. 1992, 27, 1077.
24. Perchalski, R.J.; Yost, R.A.; Wilder, B.J. *Anal. Chem*. 1983, 55, 2002.
25. Kaiser, R.E.; Cooks, R.G.; Moss, J.; Hemberger, P.H. *Rapid Commun. Mass Spectrom*. 1989, 3, 50.
26. Kaiser, R.E.; Louris, J.N.; Amy, J.W.; Cooks, R.G. *Rapid Commun. Mass Spectrom*. 1989, 3, 225.
27. Schwartz, J.C.; Syka, J.E.P.; Jardine, I. *J. Am. Soc. Mass Spectrom*. 1991, 2, 198.
28. Goeringer, D.E.; Whitten, W.B.; Ramsey, J.M.; McLuckey, S.A.; Glish, G.L. *Anal. Chem*. 1992, 64, 1434.
29. Londry, F.A.; Wells, G.J.; March, R.E. *Rapid Commun. Mass Spectrom*. 1993, 7, 43.
30. Doroshenko, V.M.; Cornish, T.J.; Cotter, R.J. *Rapid Commun. Mass Spectrom*. 1992, 6, 753.
31. Jonscher, K.; Currie, G.; McCormack, A.L.; Yates, J.R., III. *Rapid Commun. Mass Spectrom*. 1993, 7, 20.
32. Schwartz, J.C.; Bier, M.E. *Rapid Commun. Mass Spectrom*. 1993, 7, 27.
33. Stafford, G.C.; Kelley, P.E.; Syka, J.E.P.; Reynolds, W.E.; Todd, J.F.J. *Int. J. Mass Spectrom. Ion Processes*. 1984, 60, 85.
34. Vargas, R.R. Development and Application of a Laser Desorption Ionization/Quadrupole Ion Trap Mass Spectrometer. Ph.D. thesis. University of Florida, Gainesville, 1993.
35. Vargas, R.R.; Yost, R.A. The 43rd Pittsburgh Conference, New Orleans, March 9 to 13, 1992.
36. Guidugli, F.; Traldi, P. *Rapid Commun. Mass Spectrom*. 1991, 5, 343.
37. Morand, K.L.; Lammert, S.A.; Cooks, R.G. *Rapid Commun. Mass Spectrom*. 1991, 5, 491.
38. Eades, D.M.; Yost, R.A. *Rapid Commun. Mass Spectrom*. 1992, 6, 573.
39. Pedder, R.E.; Johnson, J.V.; Yost, R.A. *Proc. 41st ASMS Conf. Mass Spectrom. Allied Topics*. San Francisco, May 30 to June 4, 1993.
40. Chambers, D.M.; Goeringer, D.E.; McLuckey, S.A.; Glish, G.L. *Anal. Chem*. 1993, 65, 14.
41. Vargas, R.R.; Yost, R.A.; Lee, M.S.; Moon, S.L.; Rosenberg, I.E. 44th Pittsburgh Conf., Atlanta, March 8 to 12, 1993.
42. Amster, I.J.; Land, D.P.; Hemminger, J.C.; McIver, R.T., Jr. *Anal. Chem*. 1989, 61, 184.

PART 5

Ion Traps in the Study of Physics

Chapter 7

HIGH PRECISION MASS SPECTROMETRY IN THE PENNING TRAP

Fernande Vedel and Günther Werth

CONTENTS

I. Introduction 238

II. General Properties of Penning Traps 239
    A. Equation of Ion Motion 239
    B. Characteristic Frequencies 240

III. Practical Realization of Penning Traps 243
    A. Trap Imperfections and Frequency Shifts 243
        1. Electrode Imperfections 243
        2. Magnetic Field Gradients 244
        3. Misalignments 246
        4. Space Charge 246
        5. Magnetic Field Fluctuations 247
    B. Different Trap Geometries 248
        1. Cylindrical Traps 248
        2. Ion Cyclotron Resonance Cells 248
    C. Ion Cooling 249
        1. Adiabatic Amplitude Reduction 249
        2. Resistive Cooling 250
        3. Sideband Cooling 250
        4. Radiation Damping 251

0-8493-4452-2/95/$0.00+$.50
© 1995 by CRC Press, Inc.

5.   Buffer Gas Cooling 251
6.   Stochastic Cooling 252

IV.   Detection of Resonances 252
A.   Nondestructive Detection 252
B.   Destructive Detection 253

V.   Results 255

VI.   Conclusion 258

References 261

INTRODUCTION

Mass resolution on the order of 10^6 is required for unambiguous identification of molecules involved in chemical reactions. Mass doublets such as $^{15}NH_3$ and $^{14}NDH_2$, for example, differ by about five parts in 10^4, while the masses of CO and N_2 differ by some four parts in 10^4.

Resolutions beyond that value, in particular at the 10^9 level, are of basic importance in different fields of physics. For comparison of precise atomic physics measurements with theoretical calculations such as the Rydberg constant or Lambshift measurements, accurate values of elementary constants are needed, in particular, the mass ratio of protons to electrons. A determination of the neutrino mass from the beta decay of ^3H requires the knowledge of the mass difference of ^3He and ^3H. A comparison of masses of particles and antiparticles tests fundamental symmetry principles and, finally, mass values of chains of isotopes of one element may lead to improved models of atomic nuclei.

One way to approach and possibly exceed the 10^9 barrier in resolution is mass spectroscopy (MS) in Penning traps. Here, ions are confined by static electric fields and a superimposed magnetic field as first described by Penning.[1] The device as shown in Figure 1 has some favorable intrinsic properties which makes it extremely well suited for high precision measurements: the storage volume is on the order of a few cubic millimeters or less; the motion of the ion inside the trap can be well described by classical electrodynamics for a given electrode configuration; the observation time can be made arbitrarily long; perturbations by collisions with background gas atoms can be made very small under ultrahigh vacuum conditions; and the sensitivity of ion detection is high enough to work with single particles which avoids perturbing effects of space

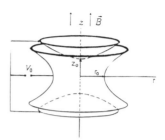

FIGURE 1
Schematic diagram of a Penning ion trap.

charge. The small volume occupied by the ions greatly facilitates the creation of both a magnetic field as homogeneous as possible and a near-perfect electric potential over the entire trajectory of an ion.

The use of the Penning trap as a mass spectrometer is based on the fact that the motion of the ions inside the trapping fields contains frequencies which depend on the particle mass. Most noteworthy is the cyclotron frequency $\omega_c = eB/m$, which is slightly changed by the presence of the electric field. An excitation of this ion motion component by radiofrequency (RF) fields can be observed and the ratio of cyclotron frequencies of different ions directly gives the ratio of their masses. Depending on the experimental method chosen for the detection of the cyclotron resonances, the mass range can be arbitrarily wide so long as the trapped ions move within the stable range of a Penning trap; the range is given by the strengths of the applied electric and magnetic fields.

As is shown below, possible systematic shifts of this frequency due to imperfections of the trapping field can be corrected to a high degree. The resolution depends finally on the coherent interaction time of the ions with the RF field, which can be made as long as several seconds. The resulting linewidth is <1 Hz at transition frequencies of several tens of megahertz in strong magnetic fields. The possibility of splitting the lines to a certain extent, depending on the signal-to-noise (S/N) ratio, has led to several experiments, which demonstrate the possibility of mass comparisons at the 1 ppb level.

II. GENERAL PROPERTIES OF PENNING TRAPS

A. Equation of Ion Motion

In order to bind harmonically an ion with charge q, the applied force should be attractive and proportional to the distance from the center

$$\bar{F} = q\bar{E} \propto -\bar{r}$$

(1)

The corresponding potential consequently must be quadratic in r including rotational symmetry around the z-axis. The simplest solution of the Laplace equation is

$$\Phi = \text{const} \left(x^2 + y^2 - 2z^2\right) \tag{2}$$

which yields the well-known electrode configuration of two hyperboloids of revolution. The constant is given by $U / 2r_0^2$, where U is the voltage applied between the ring and the end-cap electrodes and r_0 is the radius of the trap electrode (Figure 1). When the end-cap electrodes are at ground potential, we have

$$\Phi = \frac{U}{2} - \frac{U}{2r_0^2}\left(x^2 + y^2 - 2z^2\right) \tag{3}$$

From Equation 2 it is obvious that for electrostatic fields the force cannot be binding in all three directions.

In the Penning trap, the sign of the voltage is chosen in order to obtain an harmonic potential along the z-direction. The defocusing of the motion in the radial direction is avoided thanks to the Lorentz force induced by the presence of an homogeneous magnetic field in the z-direction.

B. Characteristic Frequencies

The equation of motion of a charged particle in the above-described combination of electric and magnetic fields

$$m\ddot{\mathbf{r}} = q\bar{\mathbf{v}} \times \bar{\mathbf{B}} + q\bar{\mathbf{E}} \tag{4}$$

leads to three differential equations. For the z-direction

$$\ddot{z}(t) + \omega_z z(t) = 0 \tag{5}$$

with

$$\omega_z = \sqrt{\frac{2eU}{mr_0^2}} \tag{6}$$

For the x- or y-direction

$$\ddot{x}(t) - \omega_c \dot{y}(t) - \frac{\omega_z^2}{2} x(t) = 0 \tag{7}$$

$$\ddot{y}(t) - \omega_c \dot{x}(t) - \frac{\omega_z^2}{2} y(t) = 0 \tag{8}$$

$$\omega_c = \frac{e}{m}|\mathbf{B}| \tag{9}$$

The z-component follows the motion of an harmonic oscillator. The solution of Equations 7 and 8 can be found by introducing a new variable

$$u = x + iy \tag{10}$$

such that Equations 7 and 8 give

$$\ddot{u}(t) + i\omega_c \dot{u}(t) - \frac{\omega_z^2}{2} u(t) = 0 \tag{11}$$

the solution of which is

$$u = u_0 \exp(-i\omega t) \tag{12}$$

where ω is given by

$$\omega = \frac{\omega_c}{2} \pm \sqrt{\frac{\omega_c^2}{4} - \frac{\omega_z^2}{2}} \tag{13}$$

while the perturbed cyclotron frequency is given by

$$\omega_c' = \frac{\omega_c}{2} + \sqrt{\frac{\omega_c^2}{4} - \frac{\omega_z^2}{2}} \tag{14}$$

and the magnetron frequency by

$$\omega_m = \frac{\omega_c}{2} - \sqrt{\frac{\omega_c^2}{4} - \frac{\omega_z^2}{2}} \tag{15}$$

The superposition of the two radial motions gives the form of an epicycloid (Figure 2): in the presence of the electric field the cyclotron fre-

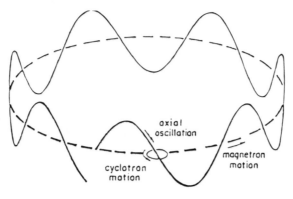

FIGURE 2
Ion trajectory in an ideal Penning ion trap.

quency, ω'_c, of the particle is shifted slightly away from the free particle value, ω_c, while the magnetron motion represents a drift of the cyclotron orbit center around the trap axis with frequency ω_m. From the above equations, we derive a criterion for stable motion from the fact that the value under the square root must be positive

$$\frac{\omega_c^2}{4} - \frac{\omega_z^2}{2} > 0 \tag{16}$$

which leads to

$$B^2 > \frac{4mU}{er_0^2} \tag{17}$$

or

$$\frac{m}{e} < \frac{B^2 r_0^2}{4U} \tag{18}$$

The stability criterion represents for given values of U and B an upper limit for the mass that can be stabilized. All ions of smaller mass can be confined simultaneously. If the stability criterion is fulfilled we have between the eigenfrequencies the relation

$$\omega'_c \gg \omega_z \gg \omega_m \tag{19}$$

In this approximation

$$\omega_m = \frac{\omega_z^2}{2\omega'_c} \tag{20}$$

is independent of the particle mass.

Between the three different eigenfrequencies we have two important relations

$$\omega_c = \omega'_c + \omega_m \tag{21}$$

and

$$\omega_c^2 = \omega'^2_c + \omega_m^2 + \omega_z^2 \tag{22}$$

Typical values for these frequencies in a trap of 1 cm radius, a direct current (DC) potential of a few volts, a magnetic field of 5 T, and light ions (protons, He$^+$, etc.) are

$$\omega_m \approx 2\pi \times 10 \ \text{kHz}$$
$$\omega_z \approx 2\pi \times 1 \ \text{MHz}$$
$$\omega'_c \approx 2\pi \times (10 \ \text{to} \ 100) \ \text{MHz}$$

III. PRACTICAL REALIZATION OF PENNING TRAPS

In our discussion we restrict ourselves to a comparison of those spectrometers that have already yielded mass measurements in the parts per billion range. These instruments are located at the University of Washington (UW),[2] the Massachusetts Institute of Technology (MIT),[3] and the University of Mainz (MZ).[4] In all cases the material for the trap is OFHC copper, gold plated to avoid surface charges. Additional important trap parameters are listed in Table 1.

A. Trap Imperfections and Frequency Shifts

A real trap always deviates from the ideal case to some extent. Such deviations will shift the eigenfrequencies of trapped ions and thus may affect the measurement of their masses. Moreover, the presence of many ions inside the trap creates a space-charge potential, which in turn may influence the eigenfrequencies. The precision of a device depends on the degree to which these imperfections can be compensated.

1. Electrode Imperfections

Electrode imperfections may arise from the truncation of the electrodes, deviations of the electrode surfaces from the hyperbolic form, and misalignment of the electrode structures; each of these imperfections can lead to anharmonicities in the potential. When the potential is expanded as a power series in spherical coordinates

$$\frac{\Phi}{U} = \sum_{k=0}^{\infty} C_k \left(\frac{r}{r_0}\right)^k P_k(\cos \delta) \tag{23}$$

TABLE 1

Parameters for the Ion Traps at UW, MIT, and MZ

Trap parameter	UW	MIT	MZ
Trap radius	2 mm	7 mm	6 mm
Applied electric potential	20–80 V	0.2–0.5 V	0.5–10 V
Magnetic field	5–6 T	8.5 T	5.8 T
Trap temp	4.2 K	4.2 K	300 K
C_4(anharmonic potential coefficient)	$\leq 1 \times 10^{-5}$	$\leq 5 \times 10^{-5}$	$\approx 10^{-5}$
B_2 (quadratic magnetic field gradient)	$2–4 \times 10^{-5}$	1.2×10^{-6}	2×10^{-6}

nonvanishing coefficients (apart from C_2, which represents the ideal potential) give rise to frequency shifts. When symmetry upon reflection $z \to -z$ is retained only even terms in the expansion contribute. Usually the largest shifts arise from C_4

$$\frac{\delta\omega_z}{\omega_z} = \frac{3}{2}C_4\left(\bar{z}/r_0\right)^2 \tag{24}$$

$$\frac{\delta\omega_m}{\omega_m} \approx \frac{3}{2}C_4\left(2\bar{z}^2 - \bar{r}_m^2\right)/r_0^2 \tag{25}$$

$$\frac{\delta\omega_c'}{\omega_c'} \approx 3C_4\frac{\omega_m}{\omega_c'}\left(\bar{z}^2 - \bar{r}_m^2\right)/r_0^2 \tag{26}$$

where $\bar{z}$ and $\bar{r}_m$ are the average amplitudes of a trapped ion in the axial and magnetron oscillations, respectively. It is assumed that the cyclotron

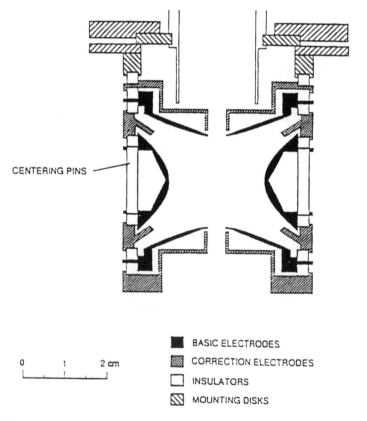

CENTERING PINS

	BASIC ELECTRODES
	CORRECTION ELECTRODES
	INSULATORS
	MOUNTING DISKS

0   1   2 cm

FIGURE 3

Penning trap showing correction electrodes for compensation of electric trapping-field imperfections.

radius r_c is much smaller than $\bar{z}$ and $\bar{r}_m$ and, therefore, terms containing r_c are neglected.

The coefficient C_4 can be minimized by additional compensation electrodes (Figure 3), placed between the ring and the end-cap electrodes, and the potential can be chosen to correct to a certain extent the anharmonicities. A very precise approach to this problem has been carried out in Seattle and is reported elsewhere.[5] The expansion coefficients of the solutions of Laplace's equation are evaluated for a variety of configurations of compensated Penning traps having hyperbolic ring and end-cap electrodes. The coefficients are calculated as a function of the location of the compensation electrodes, the shape of these electrodes, and as a function of the choice of the hyperbolas. It is shown that the coefficients are severely modified by holes and slits in the hyperbolic electrodes and also by imperfections and misalignments. The optimal asymptotes are demonstrated to lie between the symmetric trap ($r_0 = \sqrt{2}z_0$) and ($r_0 = z_0$). Because the potential of a compensation electrode is largely screened by the end-cap electrodes and the ring electrode, the shape of the compensation potential in the center of the trap is almost independent of the shape and the location of the compensation electrode. Adjustment of the compensation potential (to tune out anharmonicities) does not theoretically change the harmonic-oscillation frequency ω_z, thus, one is encouraged in the use of this frequency for high precision measurements. Values of C_4 which are actually obtained in different apparatus, are listed in Table 1.

2. Magnetic Field Gradients

While linear magnetic field gradients are cancelled by averaging over trapped ion harmonic oscillations, a quadratic variation

$$B = B_0\left(1 + B_2\left(z/r_0\right)^2\right) \tag{27}$$

leads to frequency shifts[4,6]

$$\frac{\delta\omega_z}{\omega_z} = 0 \tag{28}$$

$$\frac{\delta\omega_m}{\omega_m} = -\frac{1}{2}B_2\left(\bar{z}^2 - \bar{r}_m^2\right)/r_0^2 \tag{29}$$

$$\frac{\delta\omega_c'}{\omega_c'} = \frac{1}{2}B_2\left(\bar{z}^2 - \bar{r}_m^2\right)/r_0^2 \tag{30}$$

Commercially available superconducting solenoids with a diameter-to-length ratio of 0.1 have inhomogeneities $<10^{-6}$ in a volume of about

1 cm^3. Shimming coils, which produce additional fields having a squared dependence on the coordinates, may be used to make the quadratic coefficient B_2, which is of importance here, very small. Experimentally obtained values are listed in Table 1.

3. Misalignments

In the derivation of the ions' eigenfrequencies in a perfect trap, it is assumed that the symmetry axis of the ion trap coincides with the direction of the magnetic field. When these two directions are inclined to each other at an angle θ, the frequencies are shifted[4,6] as

$$\overline{\omega}'_c = \omega'_c + \frac{3}{2}\omega_m \sin^2 \theta \tag{31}$$

$$\overline{\omega}_z = \omega_z - \frac{3}{4}\omega_z \sin^2 \theta \tag{32}$$

$$\overline{\omega}_m = \omega_m + \frac{3}{4}\omega_m \sin^2\theta \tag{33}$$

$$\omega_c^2 = \overline{\omega}'^2_c + \overline{\omega}^2_z + \overline{\omega}^2_m \tag{34}$$

Here $\overline{\omega}'_c$, $\overline{\omega}_z$, and $\overline{\omega}_m$ are the experimentally measured frequencies and ω'_c, ω_z, and ω_m represent the ideal case. Brown and Gabrielse,[6] however, have shown that the relation (34) is, to a first-order approximation, independent of the angle θ.

The measurement of all three frequencies would eliminate this uncertainty. Because $\omega'_c \gg \omega_z$ and ω_m, the required precision in ω_z and ω_m is much less than in ω'_c to reach the 1 ppb level of uncertainties.

4. Space Charge

The shift of the trapped ions' eigenfrequencies due to the space-charge field depends on the form of the ion cloud and can vary according to the experimental conditions. When two or more ions of the same species are trapped simultaneously, shifts of the most important frequencies, ω'_c or ω_c, can be observed. Gerz et al.[4] have measured a shift in the cyclotron frequency ω_c of about 10^{-8} per additional ion; they failed to observe a shift in the perturbed cyclotron frequency ω'_c. In experiments performed at UW[7] no shift of ω_c was observed, while ω'_c was shifted by about 0.5×10^{-9} per additional ion (Figure 4). As a consequence, there is a tendency to work with single trapped ions, in which no shift occurs. Ions of different masses, which also may be trapped simultaneously and would shift the eigenfrequencies, can be eliminated by a strong resonant excitation of their z-oscillation amplitude. This drives them out of the trap.

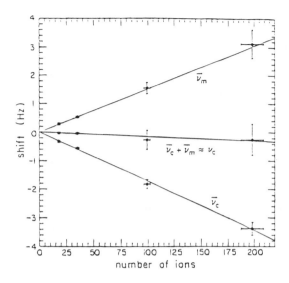

FIGURE 4

Number dependence of eigenfrequencies of trapped protons in a Penning trap. (From Van Dyck, R.S., et al., *Phys. Rev.* 1989, A40, 6308. With permission.)

On the other hand, the simultaneous storage of two ions, whose mass difference is to be measured, would circumvent the problem of temporal instability of the magnetic field (see below). Cornell et al.[8] have pointed out that when the interion spacing of two ions is held constant, cyclotron frequency perturbations from ion/ion repulsion are limited, and that ions can be brought into orbits in which measurement errors are minimized.

5. Magnetic Field Fluctuations

Magnetic fields of superconducting solenoids have been shown to fluctuate and drift due to flux jumps in the superconducting material or temperature changes in the cryostat, which might modify the susceptibility of the magnet material. Present-day technology leads to drifts on the order of 10^{-9}/h, but short-term fluctuations can be larger. Moreover, varying fields from outside will change the B-field at the trap position, even though these varying fields are shielded by the induced currents in the superconducting solenoid which are about one order of magnitude larger. Gabrielse and Tan[9] describe a self-shielding solenoid having a special geometry that reduces field fluctuations from outside by a factor of ca. 150, and thus leads to higher stability. While the electric trap parameters can be sufficiently well controlled, the stability in time of the magnetic field is probably the limiting factor in the final precision that may be obtained in Penning ion trap mass spectrometry (ITMS).

B. Different Trap Geometries

In addition to the hyperbolically shaped electrode configurations, one may attempt both to confine ions in ion traps having simpler electrode geometries and to measure their frequencies of motion so as to obtain precise mass determinations. Two attempts are discussed here.

1. Cylindrical Traps

A cylinder, closed by isolated flat end-cap electrodes creates an electrostatic potential, which near the center of the device is almost identical to the ideal quadrupolar potential. Deviations, expressed by the expansion (Equation 23) now contain large amplitudes of the higher perturbing coefficients. They can, however, be minimized by the introduction of additional compensating electrodes between the ring and end-cap electrodes (Figure 5a). Tan and Gabrielse[10] have calculated that a height of the compensating electrodes corresponding to 20% of the ring radius is adequate to avoid an unacceptable sensitivity to mechanical tolerances. The advantage of this device is that the electrodes are much easier to machine and to adjust than are hyperbolic electrodes. Experimental results[10] show that values of the compensating electrode potential, calculated by

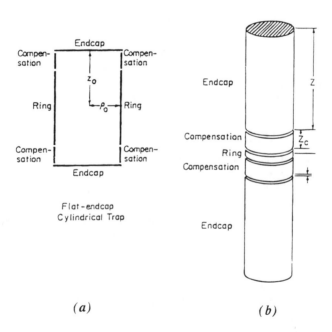

(a)   (b)

FIGURE 5

Cylindrical traps with (a) flat end-cap electrodes, (b) or open cylinders as end-caps. (From Tan, J., Gabrielse, G., *Appl. Phys. Lett.*, 1989, 55, 2144. With permission.)

standard methods of electrodynamics to minimize the higher-order co-efficients, C_k, agree well with experiment and that the leading term C_4 can be made as small as 10^{-5}.

A variant of this trap uses open cylinders as end-cap electrodes (Figure 5b). This allows easy access to the interior, which is important if particles generated externally are to be injected into the ion trap. Gabrielse et al.[11] have shown that for the case of $z_c/z_0 = 0.835$, where z_c is the length of the compensating electrode and z_0 the distance from the center to the end-cap cylinder, properly chosen compensating voltages lead to a nonzero value for the anharmonic coefficient, C_4, but also for the next order anharmonic coefficient, C_6.

2. Ion Cyclotron Resonance Cells

Apart from the hyperbolic electrode configuration in Penning ion traps, one may use mutual perpendicular parallel plates to form an electrostatic potential and to achieve trapping in a strong, superimposed magnetic field. The advantages are, obviously, the ease of machining and adjustment compared to the hyperbolic shape; there is the added possibility for adjustment of the size and shape according to the needs of the experiment. The potential, however, deviates strongly from the quadrupolar, which leads to a dependence of the eigenfrequencies of the stored ions on the position within the trap. The resonances are shifted and a precise mass determination is difficult. The resolution given by $v/dv_{1/2}$, where v is the frequency and $dv_{1/2}$ is the full width at half maximum (FWHM) of the perturbed cyclotron frequency, can be as high as 10^8.[12] Thus, the instrument functions very well for ion identification such as is required in chemical reactions, photodissociation research, and the determination of molecular structures. A number of review articles have appeared recently that describe the present status of this technique as well as that of other geometrical configurations such as hybrid combinations of hyperbolic and rectangular electrodes.[13–16]

C. Ion Cooling

As evident from Equations 24 to 30, the frequency shifts that arise from trap imperfections vary with the square of the radii of the different motions, in particular, with the amplitude of the axial and magnetron oscillations, because the cyclotron orbit of trapped ions usually is very small. In order to reduce these frequency shifts, ion motion in each of the different degrees of freedom must be cooled.

1. Adiabatic Amplitude Reduction

Ions are confined initially at a low DC trap voltage. When the voltage is raised at a rate which is given by

$$\frac{dU(t)}{dt} < \omega_z U(t) \qquad (35)$$

the z-amplitudes of the ions are reduced, while, at the same time, the kinetic energies of the ions increase such that

$$\omega_z z^2 = \text{const} \qquad (36)$$

Because $\omega_z(t) \propto \sqrt{U(t)}$, the amplitudes of the ions are reduced by the factor $(U_i/U_f)^{1/4}$, where U_i and U_f are the initial and final values, respectively, of the trapping voltage.

2. Resistive Cooling

The oscillating ion induces image charges in the end-cap electrodes of the trap which, in turn, cause an oscillating current to flow through a resistor R, connecting the two end-caps. Energy loss in this resistor dampens the axial oscillation. The damping time constant has been calculated as[17,18]

$$\tau = \left[\frac{2z_0^2}{e}\right]\frac{m}{R} \qquad (37)$$

The resistance R may be realized by a tuned circuit, which uses the trap electrodes as capacitance and an inductance to connect the end-cap electrodes. Although the time constants may be quite long (25 s for $z_0 = 1$ cm, $R = 1$ M Ω and unit mass), the method can be successfully applied in experiments that use nondestructive ion detection, because the ions are kept inside the trap essentially for infinitely long times. The final ion temperature is given by the temperature of the resistance, which may be as low as 4.2 K, if it is immersed in a liquid helium bath.[18]

3. Sideband Cooling

Transitions that induce amplitude changes in two independent ion oscillations simultaneously couple these oscillations and, therefore, may be used to transfer energy from one mode into the other.[19] When the ions are subject to a frequency $\omega_z + \omega_m$ and when the axial motion is damped by resistive cooling, as described in the previous paragraph, the magnetron radius will be reduced at the same rate as the z-oscillation amplitude. Similarly, the magnetron motion can be cooled when it is coupled to the cyclotron mode by oscillating fields at the frequency $\omega_c' + \omega_n$ and when the cyclotron orbits are reduced by radiation damping as in the case of electrons.

4. Radiation Damping

The energy of a charged particle oscillating in an trap at a frequency ω will be reduced by the emitted radiation. The loss is exponential $E = E_0 \exp(-\gamma t)$, and the damping constant γ is

$$\gamma = \frac{4e^2\omega^2}{3me^3} \tag{38}$$

Only damping constants on the order of 1 Hz or larger are of interest for MS because otherwise, the time scale becomes exceedingly long. Those time constants are obtained only for the cyclotron motion, as the other oscillation frequencies for very light particles (e.g., electrons in the Penning trap) are much smaller. For a 6-T field, the calculated damping constant is $10 \, s^{-1}$.

5. Buffer Gas Cooling

In general the use of buffer gas is prohibitive in Penning traps because the magnetron motion is metastable and any collision tends to move ions outward until they are lost from the trap. This mechanism is in contrast to Paul traps, in which collisions of the ions with light buffer gases can dampen the ion oscillations and thus lead to ion cooling and longer confinement times. It has been shown in the Penning trap, however,[20] that one can counteract the outward radial diffusion by an azimuthal quadrupole RF field oscillating at the sum-frequency of the magnetron and cyclotron motions. Then the magnetron energy is transformed into cyclotron energy and vice versa. A small amount of buffer gas will cool the axial and cyclotron motions. By adjustment of the strength of the applied RF field so that the rate of transformation is commensurate with the cooling rate of the cyclotron motion, kinetic energy will be removed from all three degrees of freedom and ions will move toward the center of the ion trap (Figure 6).

6. Stochastic Cooling

When the amplitude and the phase of the oscillation of an ion in a Penning trap has been measured, a suitable electric field can be applied whose amplitude and time duration is selected in order to subtract this component from the motion. At this point, the system may start to evolve toward a new equilibrium state in which the energy again will be shared among the three different degrees of freedom. This type of cooling has been discussed theoretically,[21] and although the first results have been reported,[22] it has not thus far been applied routinely to mass measurements.

FIGURE 6
Ion trajectory in a Penning trap in the presence of a damping force proportional to velocity and an additional quadrupole field at the frequency ω'_c-ω. Both cyclotron and magnetron radii are decreased. (From Savard, G. et al., *Phys. Lett.*, 1991, 158, 247. With permission.)

IV. DETECTION OF RESONANCES

In general, it is facile to excite the different ionic resonances ω'_c, ω_z, and ω_m by RF fields applied to the ion trap electrodes. In addition, combined transitions such as $\omega'_c \pm n\omega_m$, $n = 1, 2, \ldots$ can be driven with a suitably chosen geometry of the applied RF field. In this case the ring electrode is divided into several sections and the RF field can be applied between different parts to give the desired spatial geometry. The experimental difficulty arises in the detection of such resonances, for instance, when the ion number is very small. Of particular interest, of course, are the frequencies ω'_c and $\omega'_c - \omega_m = \omega_c$ because they are used for mass determination.

Access to the ion motion inside an ion trap is given usually by the z-oscillation of the ions. A detection of ω'_c or ω_c resonance, which represents a change in the ion's motion in the x-y plane, requires some kind of coupling between the x-y-amplitude and the z-motion. Different methods have been used to provide such coupling. In experiments performed at UW and MIT, the detection is nondestructive and the ions remain in the trap during the complete measurement, while in the experimental arrangement at MZ, the ions are ejected from the ion trap after each measurement cycle.

A. Nondestructive Detection

In the nondestructive detection of ω'_c or ω_c resonances, the ions are kept continuously in the ion trap; they induce image currents in the end-cap electrodes which contain information about their ω_z frequencies as well as their oscillation amplitudes. The frequency can be measured quite accurately by high-Q resonance circuits. The noise amplitude appearing across a resistance that connects the end-cap electrodes can be measured by square-law detection. A change in the axial oscillation amplitude is observed as an increase in the noise amplitude. Operation of the circuit at liquid helium temperature reduces the thermal noise and increases the sensitivity to the point where it is possible to detect a single trapped ion.

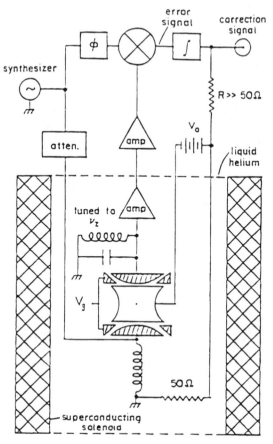

FIGURE 7

Schematic diagram of the mass spectrometer at UW. (From Van Dyck, R.S., Jr., *J. Mod. Opt.*, 39, 1992. With permission.)

The required coupling between the motion in the x-y plane and the z-direction is absent in a perfect ion trap, loaded with a single ion. Thus, additional means have to be applied to provide this coupling.

The approach of van Dyck et al.[23] at UW to provide this coupling has been to introduce a small, well-defined magnetic field inhomogeneity to the ion trap originating from a Nickel ring or a superconducting loop placed in the x-y plane at z = 0. A change in the ion cyclotron radius then induces a change in the z-oscillation amplitude which, in turn, is observed as a voltage drop across the LC circuit. Recently the group[24] has demonstrated that the remaining nonharmonic coefficients C_4 and C_6 of a non-ideal trap also give sufficient coupling between the different modes so that no extra inhomogeneity in the B-field is required. Figure 7 shows a schematic diagram of this spectrometer, including the detection of the axial resonance.

Figure 8, is an example of a single C_4^+ ion, where axial frequency shifts on the order of 0.33 Hz in a total frequency of about 5.7 MHz are clearly observed.

A different approach has been taken by the MIT group. Here, the change in the ω_c amplitude is transferred to the z-direction by a pulse applied between the ring and the end-cap electrodes; the pulse is applied at a frequency of the sideband $\omega'_c - \omega_z$, which couples both oscillation modes. A pulse with a properly chosen amplitude-duration product (π-pulse) will exchange the phase and the action of one mode with those of the other.[25] The z-oscillation amplitude then is observed via the induced image charges in a SQUID detector. Figure 9 shows a signal obtained with one or a few ions.[3]

B. Destructive Detection

In the Penning trap experiments carried out at MZ,[4] the ions are kicked out of the trap by a pulse applied to an end-cap electrode. The ions then travel along the magnetic field (Figure 10). In the inhomogeneous part of the field near the end of the solenoid, they are accelerated by a force

$$\bar{F} = \nabla(\bar{\mu} \cdot \bar{B}) \tag{39}$$

which acts upon their magnetic moment μ

$$\mu = \mu_z = \frac{e\hbar}{2mc} l_z = \frac{1}{2} e\omega_c r_c^2 \tag{40}$$

Here, l_z is the angular momentum of the ions and r_c the radius of the cyclotron orbit. Because μ_z is a constant of motion and B is directed along the z-axis, we have

$$\bar{F} = \bar{\mu} \cdot \nabla(\bar{B}) = \mu_z \frac{\delta B}{\delta z} \tag{41}$$

Ions of different masses have different times-of-flight (TOF) to a multi-channel-plate detector placed outside the B-field at a distance of about 40 cm from the ion trap (Figure 11). An RF excitation of an ion cyclotron frequency increases r_c, which leads to an increased force and thus to shorter TOF. The change in average flight time at varying RF frequencies can be used to detect the cyclotron resonance (Figure 12). Because the detection efficiency of the channel-plate detector is usually between 50 and 70%, the device can be operated with single ions.

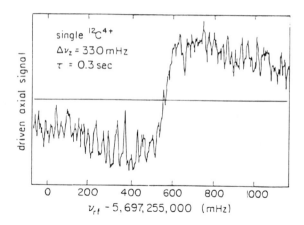

FIGURE 8
Single C_4^+ axial resonance recorded with the spectrometer at UW. (From Van Dyck, R.S., Jr., et al., *J. Mod. Opt.*, 1992, 39, 243. With permission.)

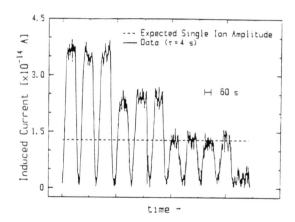

FIGURE 9
Detection signal of one, two, and three ions recorded with the spectrometer at the MIT. (From Cornell, E.A. et al., *Phys. Rev. Lett.*, 1989, 63, 1674. With permission.)

V. RESULTS

As the cyclotron frequency of an ion in a Penning trap is inversely proportional to the mass of the particle, the highest precision can be obtained with light ions, provided the linewidth is mass independent, as is usually the case. For heavier masses, the precision can be increased, when multiply charged rather than singly charged ions, states are trapped.

The setups described in the previous paragraphs at UW, MIT, and MZ have yielded a number of results on light masses and have demon-

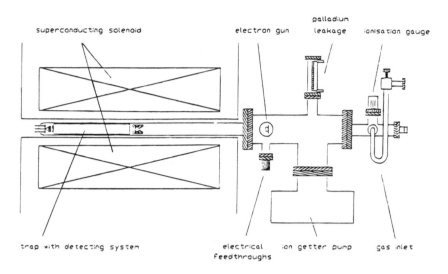

FIGURE 10

Penning trap mass spectrometer with TOF detection of ions at MZ. (From Gerz, Ch., Wilsdorf, D., Werth, G., *Nucl. Instr. Methods B.*, 1990, 47, 453.)

strated precisions in the parts per billion range for the comparison of different masses. The smallest uncertainties are obtained in mass doublets. Here, the frequency shifts cancel to a high degree. Table 2[3,26–32] gives examples of results obtained thus far for small masses. The interest in these values arises mainly from fundamental aspects. The mass values of protons, electrons, and neutrons (obtained with the help of binding energies of small nuclei) appear in basic formulas in atomic physics such as the Rydberg constant or the Lambshift in hydrogen, and are needed to a high precision in order to compare experiments and theoretical calculations.

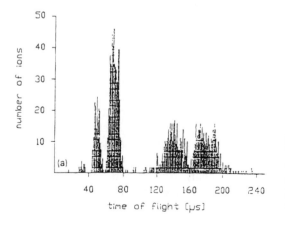

FIGURE 11

TOF spectrum of different trapped ions. (From Gerz, Ch., Wilsdorf, D., Werth, G., *Nucl. Instr. Methods B.*, 1990, 47, 453.)

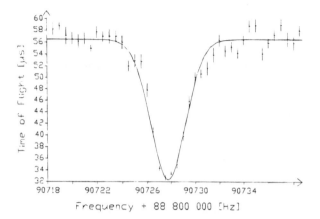

FIGURE 12

Cyclotron frequency ω_c of protons in a magnetic field of 5.8 T detected by change in TOF. The statistical uncertainty of the line center is 0.37 ppb.

TABLE 2

Precision Mass Values Obtained in Penning Traps

Particles	Ratio or difference	Relative uncertainty (ppb)		Ref.
$m_{\bar{e}}/m_o$	1.000 000 000	(130)	130	26
$M_{\bar{p}}/M_p$	0.999 999 877	(42)	42	27
M_p/m_e	1836.152 701	(37)	21	28
$M(^{12}C^{4+})/4m_p$	2.977 783 713	(10)	3	29
$M(^{12}C^{4+})/2M(^4He^{2+})$	1.499 161 233	(15)	10	30
$M(^{12}C^{4+})/4M(^3He^+)$	0.994 684 341 4	(75)	7.5	30
$M(^{18}O^{6+})/6M(^3He^+)$	0.883 862 581 6	(10)	1.1	24
$M(^{12}C^{4+})/4M(^3H^+)$	0.994 677 754	(3)	3	24
$M(^3He^+)/M(H_2^+)$	1.496 441 095	(6)	4	31
$M(D_2) - M(^4He)$	1.025 600 331	(5)	1.3	32
$M(CO^+)/M(N_2^+)$	0.999 598 887 6	(4)	0.4	3

The mass difference of ^{3}He and ^{3}H is an important quantity in the search for a finite neutrino mass from beta-decay of tritium, and a mass comparison of proton/antiproton and electron/positron is a test of the fundamental symmetry under space and time reversal.

Although the mass values of heavy nuclei are somewhat less precise, Penning trap experiments have yielded a number of results on chains of unstable isotopes of several elements, which are an improvement on the accepted mass values. These values are of interest to test nuclear models, which predict masses of unknown isotopes away from the stable region. The experiments performed by a group from MZ at the isotope facility ISOLDE at CERN[33,34] employ a tandem Penning trap arrangement (Figure 13): ions leaving the ISOLDE mass separator at some 50 keV energy are collected on a rhenium metal foil, then re-evaporated at thermal energy. They are confined initially in a large-sized trap and then transferred by an electric pulse into a smaller high-precision trap located at the most homogeneous part of the magnetic field. This final transfer is very efficient, with up to 70 percent of the ejected ions being captured successfully in the second trap. The results of these experiments have shown discrepancies in earlier measurements in which conventional mass spectrometers were used. While in traditional instruments the resolution decreases for short-lived isotopes, the Penning trap results are almost independent of the lifetime of the particles, provided the lifetime is longer than the time needed to excite the cyclotron orbit (Figure 14). The resolution could be increased to values approaching 10^7, sufficient not only to increase substantially our knowledge of atomic masses outside the range of stable isotopes, but also to resolve, for the first time, isomeric states of nuclei[34] which differ in mass from that of the ground state (Figure 15).

VI. CONCLUSION

The benign environment offered by Penning traps for the confinement of ions, with respect to the coherent observation of time and freedom from perturbations from the outside world, has been used to determine mass ratios of ions with uncertainties of 1 ppb or less. These

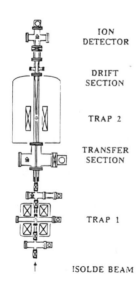

ION
DETECTOR

DRIFT
SECTION

TRAP 2

TRANSFER
SECTION

TRAP 1

ISOLDE BEAM

FIGURE 13

Layout of the tandem Penning trap mass spectrometer at ISOLDE-CERN. (From Kluge, H.-J.; Bollen, G. *Nucl. Instr. Methods.*, 1992, B70, 473. With permission.)

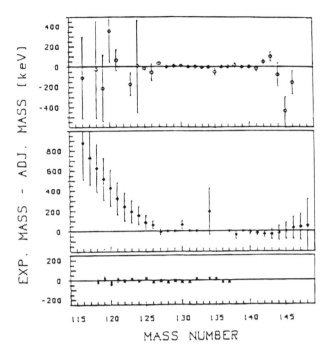

FIGURE 14
Mass data of Cs isotopes. The zero lines are the results of an adjustment of all the mass data available. The deviations from the adjusted values are plotted for the available Q_β and reaction data (top), for the results obtained by a Mattauch-Herzog mass spectrometer (middle), and for the data obtained by the Penning trap mass spectrometer with ^{133}Cs as reference isotope (bottom). (From Kluge, H.-J.; Bollen, G. *Nucl. Instr. Methods.*, 1992, B70, 473. With permission.)

achievements have already influenced the precision of fundamental constants. It is envisaged that the resolution may be increased further to the 10^{-11} level. At this stage, electronic excitation of ions would become visible by a change in the mass of the ion. Excited state particles would be distinguished from those in the ground state by MS methods, as is already the case for excited nuclear states. A further application of very high mass resolution in Penning traps may be the verification of multi-electron atomic theories such as Hartree-Fock, by weighing the bare nucleus of a high Z atom and then weighing the ground state energies as electrons are added. A special case would be the ground state of hydrogen-like ions of high Z. A mass determination could afford the means to test the theory of quantum electrodynamics in strong fields.

The mass measurements of unstable isotopes in Penning traps have already led to corrections of mass values which had been obtained previously by conventional mass spectrometers. The impact of the probable

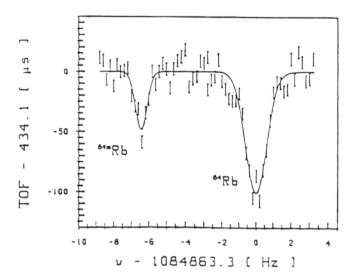

FIGURE 15
Cyclotron resonances for ⁸⁴Rb in its ground and isomeric state. (From Kluge, H.-J.; Bollen, G. *Nucl. Instr. Methods.*, 1992, B70, 473. With permission.)

improvement of nuclear models, which predict masses of unknown isotopes, is expected to be substantial.

Proposals exist[35] and active work is presently being pursued to use Penning ion traps for a new and more accurate determination of the kilogram standard. While fundamental quantities of physics, such as the time and the length, are now defined in terms of a chosen atomic standard frequency, there is not yet an agreed atomic standard of mass. The kilogram is the last basic unit represented by a manufactured object and all mass measurements must be always done by comparison with a reference, the prototype kilogram in the Bureau International des Poids et Mesures at Sèvres, France. Recently, an absolute atomic definition of mass was proposed, on the basis of the de Broglie frequency, measured as

$$m \equiv \frac{c^2}{\bar{\lambda}\gamma v} \tag{42}$$

where $\bar{\lambda}$ is the mean de Broglie wavelength of the particle moving with the velocity v, which leads to the Lorentz factor $\gamma = 1/(1 - v^2/c^2)^{1/2}$; this proposal has the advantage that mass is defined only in terms of kinematic quantities (time and length), which are directly referred to the SI standard and measured with great accuracy. A different proposal[36] would define another kilogram prototype. Experiments are in progress for using the mass of the ²⁸Si atom as an atomic mass unit and to relate this microscopic quantity to the macroscopic by a more accurate value for

Avogadro's constant.[36,37] In order to tie the present microscopic mass unit, i.e., $1/12$ of the mass of ^{12}C with that of ^{28}Si, the mass ratio of these two isotopes must be measured with an accuracy better than 10^{-8}. Such an accuracy can be achieved by doublet MS in a Penning trap: either cluster ions $^{28}Si_3{}^+/{}^{12}C_7^+$ with $m/q = 84$[38] or highly charged isobars ($^{28}Si_3{}^+/{}^{12}C_7^+$) with $q/m = 4$[39] can be considered.

REFERENCES

1. Penning, F.M. *Physica*. 1936, 3, 873.
2. Van Dyck, R.S., Jr.; Farnham, D.L.; Schwinberg, P.B. *J. Mod. Opt.* 1992, 39, 243.
3. Cornell, E.A.; Weisskoff, M.; Boyce, K.R.; Flanagan, R.W., Jr.; Lafyatis, G.P.; Pritchard, D.E. *Phys. Rev. Lett.* 1989, 63, 1674.
4. Gerz, Ch.; Wilsdorf, D.; Werth, G. *Nucl. Instr. Methods B.* 1990, 47, 453.
5. Gabrielse, G. *Phys. Rev.* 1983, A27, 2277.
6. Brown, L.S.; Gabrielse, G. *Rev. Mod. Phys.* 1986, 58, 233.
7. Van Dyck, R.S., Jr.; Moore, F.L.; Farnham, D.L.; Schwinberg, P.B. *Phys. Rev.* 1989, A40, 6308.
8. Cornell, E.A.; Boyce, K.R.; Fygenson, L.K.; Pritchard, D.E. *Phys. Rev.* 1992, A45, 3049.
9. Gabrielse, G.; Tan, J. *J. Appl. Phys.* 1988, 63, 5143.
10. Tan, J.; Gabrielse, G. *Appl. Phys. Lett.* 1989, 55, 2144.
11. Gabrielse, G.; Haarsma, L.; Rolston, X. *Int. J. Mass Spectrom. Ion Processes.* 1989, 88, 319; 1989, 93, 121.
12. Marshall, A.G.; Schweikhard, L. *Int. J. Mass Spectrom. Ion Processes.* 1992, 118/119, 37.
13. Asamoto, B.; Dunbar, R.C. *Analytical Application of Fourier Transform Ion Cyclotron Resonance Mass Spectrometry.* VCH: New York, 1991.
14. Marshall, A.G. *Adv. Mass Spectrom.* 1989, 11A, 651.
15. Nibbering, N.M.M. *Adv. Mass Spectrom.* 1989, 11A, 101.
16. Wanczek, K.-P. *Int. J. Mass Spectrom. Ion Processes.* 1989, 95, 1.
17. Walls, F.L.; Dehmelt, H.G. *Phys. Rev. Lett.* 1968, 21, 127.
18. Wineland, D.J.; Dehmelt, H.G. *J. Appl. Phys.* 1975, 46, 919.
19. Wineland, D.J.; Dehmelt, H.G. *Int. J. Mass Spectrom. Ion Phys.* 1975, 16, 251.
20. Savard, G.; Becker, St.; Bollen, G.; Kluge, H.-J.; Moore, R.B.; Otto, Th.; Schweikhard, L.; Stolzenberg, H.; Wiess, U. *Phys. Lett.* 1991, 158, 247.
21. Lagomarsino, V.L.; Manuzio, G.; Testera, G. *Phys. Rev.* 1991, A44, 5173.
22. Beverina, N.; et al. *Phys. Scripta.* 1988, T22, 238.
23. Van Dyck, R.S., Jr.; Wineland, D.J.; Ekström, P.A.; Dehmelt, H.G. *Appl. Phys. Lett.* 1976, 28, 446.
24. Van Dyck, R.S., Jr. Lecture presented at AMCO-9, Bernkastel, Germany, 1992.
25. Cornell, E.A.; Weisskoff, M.; Boyce, K.R.; Pritchard, D.E. *Phys. Rev.* 1990, A41, 312.
26. Van Dyck, R.S., Jr.; Schwinberg, P.B.; Dehmelt, H.G. *Phys. Rev. Lett.* 1981, 81, 119.
27. Gabrielse, G.; et al. *Phys. Rev. Lett.* 1990, 65, 1317.
28. Van Dyck, R.S., Jr.; Moore, F.L.; Farnham, D.L.; Schwinberg, P.B. *Bull. Am. Phys. Soc.* 1986, 31, 244.
29. Moore, F.L.; Farnham, D.L.; Schwinberg, P.B.; Van Dyck, R.S., Jr. *Bull. Am. Phys. Soc.* 1989, 34, 99.
30. Moore, F.L.; Farnham, D.L.; Schwinberg, P.B.; Van Dyck, R.S., Jr. *Nucl. Instr. Methods.* 1989, B43, 425.

31. Hagena, D.; Werth, G. *Europhys. Lett.* 1991, 15, 491.
32. Gerz, Ch.; Wilsdorf, D.; Werth, G. *Z. Phys. D. Atom. Mol. Clusters.* 1990, 17, 119.
33. Bollen, G.; et al. *J. Mod. Opt.* 1992, 39, 257.
34. Bollen, G.; et al. *Phys. Scripta.* 1992, 46, 581.
35. Wignall, J.W.G. *Phys. Rev. Lett.* 1992, 68, 5.
36. Kluge, H.-J.; Bollen, G. *Nucl. Instr. Methods.* 1992, B70, 473.
37. Taylor, B.N. *IEEE Trans. Instr. Meas.* 1991, 40, 86.
38. Lindinger, M.; Becker, St.; Bollen, G.; Dasgupta, K.; Jertz, R.; Kluge, H.-J.; Schweikhard, L.; Vogel, M.; Lützenkirchen, K. *Z. Phys.* 1991, D20, 441.
39. Jertz, R.; Bollen, G.; Kluge, H.-J.; Schweikhard, L.; Stolzenberg, H.; Bergström, I.; Carlsberg, C.; Schuch, R. *Z. Phys.* 1991, D21, 179.

Chapter 8

THE PAUL TRAP AS A COLLECTION DEVICE

Robert B. Moore and M. David Lunney

CONTENTS

I. Introduction 264

II. General Phase Space Considerations of Particle
    Collections 265
    A. Action Diagrams for a Collection of Free Particles 266
    B. Action Diagrams for a Collection of Particles
        Experiencing a Uniform Force 270
    C. Particles Experiencing a Force Proportional
        to Displacement 271
        1. Beam Focusing and Defocusing 272
        2. Particles Undergoing Simple Harmonic
            Motion 275
    D. Transfer of a Single Beam Pulse into a Trap 276
        1. Axial Phase Space Considerations of Trap
            Injection 276
        2. Transverse Phase Space Considerations of
            Trap Injection 279

III. Continuous Beam Injection Into Traps 280
    A. General Phase Space Considerations of Continuous
        Trap Injection 280
    B. Practical Systems for Continuous Trap Injection 281

0-8493-4452-2/95/$0.00+$.50
© 1995 by CRC Press, Inc.

C.   Effects of the Radiofrequency Phase on Paul
     Trap Injection 283
D.   Design of Deceleration Systems for Trap Injection 284

IV. Phase Space Distributions within a Trap 285
A.   Phase Space Distributions in an Ideal Trap 285
B.   Effects of Space Charge in an Ideal Trap 287
C.   Phase Space Distributions in a Paul Trap 287

V.   Extraction of Particles from a Trap 291
A.   Slow Extraction from Traps 292
B.   Fast Extraction from Traps 293

VI. Examination of Trapped Ions by Pulsed Extraction 296

VII. Conclusions 301

References 302

I. INTRODUCTION

The Paul trap becomes a much more useful device when ion beams can be injected into them from the outside, and once ions have been collected in the trap, they can be extracted as a well-defined beam pulse. The injection of ions from an outside source allows the use of many more ionization processes for the production of a particular ion species. An example is the production of negative ions by soft interactions with a cold electron cloud, a process which is impossible within the ion trap because of the high kinetic energy of free electrons due to the radiofrequency (RF) electric field.

Also, the extraction of an ion cloud as a well-defined pulse allows closer and more careful observation of that cloud than is usually possible while the ions are confined within the trap. For example, the ions could be extracted as a small cloud for subsequent injection into a precision Penning trap mass spectrometer of the type being used for high-accuracy mass measurements on radionuclides at the ISOLDE (Isotope Separator-Off Line-Direct Entry) facility at CERN (Centre Européenne Recherches Nucléaire).[1] Another use would be to prepare a collection of ions for precision collinear laser spectroscopy,[2] permitting the observation period to be limited to the short period when the ions were in the vicinity of the detector. By such restriction of the observation period, the signal-to-noise (S/N) ratio of the observations can be enhanced considerably. This enhancement of the S/N ratio is very valuable when there are only a limited number of ions available, such as is often the case with radionuclides.

A particular use of Paul traps that is of great interest is as an accumulator for a steady stream of ions from atmospheric pressure ion (API) sources that give high yields of highly ionized biomolecules of high mass. The collected ions could then be extracted as a sharply defined pulse for time-of-flight (TOF) mass spectrometry (MS). In this way, the high S/N performance of the TOF technique could be combined with the high yields of the API sources. Very interesting results obtained with a combination of a TOF mass spectrometer and a Paul trap have already been reported.[3]

In these applications, the Paul trap is used as a beam collection and cooling device in order to prepare an ion cloud for subsequent measurements. This chapter presents some of the generalities of the transfer of beams of charged particles in and out of traps with specific reference to the Paul trap. It is assumed that the reader knows the geometry of a Paul trap and has an understanding of the basic principles of ion motion within such a trap, as discussed in Chapter 2 of Volume I. There is also some discussion of the results of present systems used for such injection and extraction. However, because of the background of the authors, this discussion is biased toward the work in physics rather than chemistry.

II. GENERAL PHASE SPACE CONSIDERATIONS OF PARTICLE COLLECTIONS

As in any transformation of a collection of particles, the concept of phase space is a very powerful tool for dealing with the transfer of ions into and out of Paul traps. This has been shown by Todd et al.,[4,5] who in 1980 presented a mathematical formalism for dealing with the phase space volume of such traps. What is presented here is a more gradual and intuitive approach to the concepts, in the hope that the formalism will become more readily understandable.

Phase space is the six-dimensional space out of which the ordinary three-dimensional space and the three-dimensional momentum space can be projected. Essentially, the position and momentum co-ordinates of each particle in a collection at any particular instant are represented by a point in this space. The significance of phase space is that the motion of the particle points in phase space describes completely the dynamics of a particle collection. Phase space is particularly appropriate for manipulations into and out of traps because the times involved in such transfers are usually so small that particle/particle interactions are insignificant compared to the effects of the applied electric fields. In such a situation, the classical theorem of Liouville applies; that throughout the transformations the phase space density of the particle collection is everywhere preserved. Essentially, in phase space the ion collection behaves like an incompressible fluid.

To understand the result of manipulations of a collection of ions into and out of a Paul trap, it is necessary, therefore, to understand phase space. However, the full six-dimensional phase space of a particle collection is difficult to visualize. Fortunately, in most of the cases of practical interest in ion transfer into and out of traps, the applied fields are designed to result in linear equations of motion which are then separable into the three displacement coordinates. In practical terms, this means that the electric field components of order higher than quadrupolar are not significant. A device in which the electric fields are purely quadrupolar is the quadrupole ion trap of Paul. With such a device, deceleration, injection, and extraction systems are free of significant aberrations.

For such fields, particular significance is assigned to the projection of the particle points of phase space onto the two-dimensional momentum-displacement planes for each coordinate as, in each such projection, the density of the points in the plane is itself preserved. Diagrams representing such two-dimensional projections may be called "action diagrams" because an elemental area in such a diagram has the units of action, i.e., momentum × displacement or, equivalently, energy × time.

In charged particle manipulation, the usual unit for the area of an action diagram is the electronvolt-second (eV-s). For the energies and times involved in charged particle manipulation in an electromagnetic trap, the most appropriate unit is the eV-μs. The most appropriate unit for momentum is eV-μs/mm. For reference, it is convenient to note that the momentum of a 1-eV ion of $m = 100$ Da is 1.440 eV-μs/mm, and increases in proportion to the square root of the atomic mass.

To show the significance of action diagrams, those for some simple collections of particles are presented below.

A. Action Diagrams for a Collection of Free Particles

The simplest possible action diagram is that for a collection of identical particles moving freely through ordinary space. Because the motions in the three spatial coordinates are independent, the behavior of the particles in the three action diagrams will be independent and any one of these action diagrams will be representative of the particle collection.

Consider, therefore, only the diagram for the motion in one of the Cartesian coordinates x and, as a further simplification, suppose that it is as shown in Figure 1; initially, the collection of particles occupies a rectangle of width Δx and height Δp_x centered on the coordinate axis, where Δx is the spread of the collection of particles in the x-direction, and Δp_x is the range of their momenta, p_x, in the x-direction. The action diagrams in this figure are for the initial collection at time $t = 0$ and for a time Δt

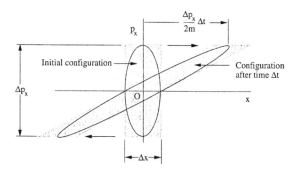

FIGURE 1

Action diagrams for a collection of identical free particles of mass m, which at time $t = 0$ fill a rectangle, and after drifting for a time Δt, fill a parallelogram. Particles on the ellipse in the initial diagram would drift to be on the ellipse shown within the parallelogram.

later. The particles are thus all located between $\pm \frac{1}{2}\Delta x$ and have momenta that range between $\pm \frac{1}{2}\Delta p_x$. In other words, the particle collection is centered on the spatial coordinate system and has zero central momentum.

At a later time, all the particles will have retained their original momentum coordinates, but their horizontal coordinates will have changed by an amount which is simply the product of their velocities and time, Δt, which is expressed in terms of p_x and the particle mass m as

$$x = \frac{p_x}{m}\Delta t \qquad (1)$$

Because x is proportional to the p_x coordinate, the points filling the original rectangle in the action diagram will move so as to fill the parallelogram shown in the figure. It can be easily seen that the area of this parallelogram is the same as that of the original rectangle. Thus, while the spatial volume of the particles actually expands due to the particle motions, the action area is conserved.

Furthermore, ellipses that have the same area can be enclosed by the two diagrams. This point is important in practical charged point manipulation, where the collection to be manipulated often falls within elliptical shapes in action diagrams (see Section II.C.2).

Similarly, the action areas are conserved for the y- and z-components of the motion. The six-dimensional phase space volume occupied by the particle collection will also remain constant during the motion.

A very similar action diagram describes a collection of particles centered about a finite momentum. The only difference from that of Figure 1 is that the parallelogram would move to the right by an amount equal to the central velocity multiplied by the time interval. Such diagrams are important in charged particle beam optics, where one usually defines the z-axis to be that of the direction of the central momentum of the particle beam. The transverse action diagrams for x and y are, therefore, like those of Figure 1. As an illustration of the use of such action diagrams in beam optics, projections of a beam of particles going through a focus in the transverse y coordinate with a central axial momentum of p_{z0} are shown in Figure 2. It can be seen that the minimum width of such a beam occurs when the action diagram is symmetric about the transverse momentum (p_y) axis. This condition is often taken to be the definition of a focus in a beam.

It is common in beam optics to express a transverse momentum of a particle as a ratio of the central axial momentum, expressed as p_{z0} where p_z is the particle momentum in the direction of the beam axis. This ratio, $\Delta p_y/p_{z0}$, defines the angular divergence of the particle from the central axis, and is usually expressed in milliradians. Action diagrams in which the transverse momenta are so expressed are called emittance diagrams, and the area that encloses the particles in such diagrams is called the beam emittance. The relation between a transverse action area, A_y, of a beam and its emittance, ξ_y is given simply as

$$A_y = p_{z0}\xi_y \tag{2}$$

An oft-used unit for emittance is the millimeter-milliradian (mm-mrad). For reference, it is convenient to note that the transverse action of a 10-keV beam of particles of $m = 100$ Da with emittance 1 mm-mrad is 0.1440 eV-μs, and increases in proportion to the emittance and to the square root of both the atomic mass and the beam energy.

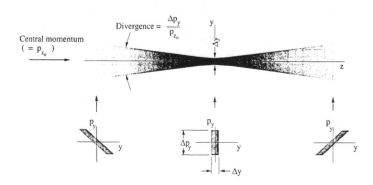

FIGURE 2

(Upper) The projection in the y-z plane of a beam of particles with central momentum p_{z0} going through a focus. The transverse momentum spread in the y-direction is taken as Δp_y and the beam width at the focus is Δy. (Lower) The action diagrams, p_y vs. y, at various points in the beam. The shaded arrows indicate the positions in the beam at which the action diagrams apply.

Axial, or longitudinal, action diagrams are used in beam optics when dealing with beam pulses. The action area for a beam pulse is normally referred to as its "longitudinal emittance" and, as already mentioned, is usually expressed in eV-μs. The significance of this unit is seen more clearly when the action diagrams are presented as the energy E of a particle plotted against the time t at which the particle arrives at a specific z coordinate along the beam axis. The relationship between such diagrams and the momentum-displacement action diagrams is shown in Figure 3 for a collection of particles with an energy spread ΔE. In this figure, the action diagrams on the far left show particles that are at the rear of the pulse having lower z (or greater t); such particles are seen to have higher momenta and, therefore, higher energy. The diagrams at the center are for a later time and further downstream when the higher energy particles have caught up with the slower ones and have achieved a "time focus". The time focus is defined to be where and when the particle collection takes a minimum time to pass. The diagrams to the right show the higher momentum particles advancing past the slower ones.

That the two sets of action diagrams in Figure 3 have the same area can be seen easily for the case in which the momentum spread is much smaller than the central momentum and the velocities are nonrelativistic. Then

$$\Delta E = v_{z0}\Delta p \qquad \Delta t = \frac{\Delta z}{v_{z0}} \qquad \Delta E \, \Delta t = \Delta p \, \Delta z \qquad (3)$$

where v_{z0} is the central velocity of the beam pulse, ΔE is the energy spread of the particle collection, and Δt is the time spread of the particle collection at the time focus.

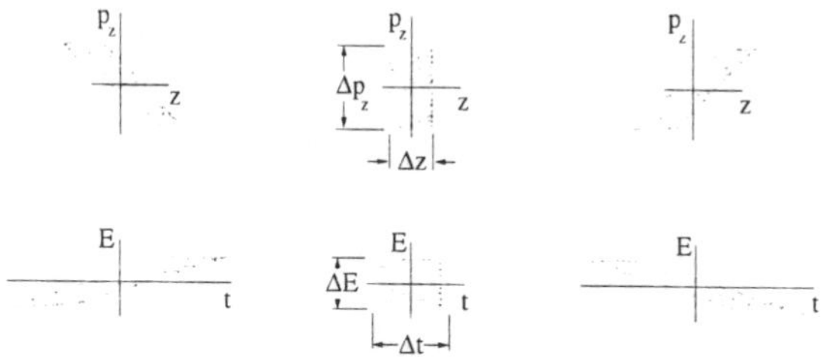

FIGURE 3
(Upper) Three momentum-displacement action diagrams for a beam pulse that is going through a time focus. The meaning of time focus is illustrated by the diagrams immediately below each momentum-displacement action diagram. These diagrams represent energy vs. TOF relative to that of the central particle of the collection. For the energy-time diagrams to have straight sides, the central momentum of the beam must be much greater than the momentum spread Δp_z of the particles.

B. Action Diagrams for a Collection of Particles Experiencing a Uniform Force

The conservation of action areas for the particle collections considered in Section II.A might appear to be a trivial consequence of having chosen free particles. However, let us consider a less trivial case. Suppose that the collection experiences a uniform force in the x-direction (see Figure 4). In this case, the change in momentum in a given time interval is the same for all the particles, and thus the points in the action diagram all move up by the same amount. The displacement coordinates all change by the amount given in Equation 1 augmented by a constant term related to the acceleration of the particles due to a force F

$$\Delta x = \frac{p_x}{m} \Delta t + \frac{F}{2m} \Delta t^2 \tag{4}$$

The action diagram after Δt will, therefore, be the same parallelogram as in Figure 1, but now moved upward and farther to the right by the constant term in Equation 4. The action area of the collection is, of course, still conserved.

The conservation of the action area of a group of accelerating particles is of great importance in the design of ion sources and particle accelerators. When the accelerating field is uniform, the action diagrams for motion along this axis are as shown in Figure 4, except that the coordinate x has been substituted for the coordinate z. Also, when the ac-

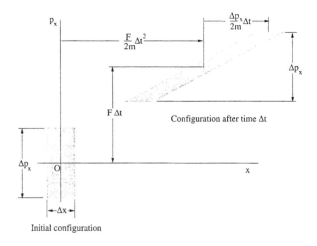

FIGURE 4
Action diagrams for a collection of identical particles of mass m, initially enclosed by a rectangle in their x action diagram, which experience a constant and uniform force F in the x-direction.

celerating field is uniform and in the z-direction, the motion in the transverse coordinates x and y is unaffected. Thus, the action diagrams for these coordinates are as shown in Figure 1.

However, if transverse momenta are expressed as divergences, the areas of the resulting emittance diagrams will change with the central momentum of the particles. To retain the concept of an emittance that is conserved, it is common in ion source and ion accelerator design to refer to a "normalized" emittance ξ_n which is defined in terms of the ordinary emittance ξ as

$$\xi_n \equiv \gamma\beta\xi \tag{5}$$

where γ and β are, respectively, the dimensionless relativistic factors of the ratio of the mass of the central beam particle to the rest mass of the particles and the ratio of the central beam velocity to the velocity of light. The relationship between a normalized emittance and an action area A is, therefore,

$$A = m_0 c \xi_n \tag{6}$$

where m_0 is the rest mass of the particles and c is the velocity of light.

It may also be useful to note that the transverse action of a beam of particles of m = 100 Da, with normalized emittance of 1 mm-mrad, is about 310 eV-μs and increases in proportion to the emittance and to the atomic mass.

C. Particles Experiencing a Force Proportional to Displacement

Proceeding to the next level of complexity, suppose that a collection of identical particles experiences a force in the x-direction which is proportional to the x displacement,

$$F_x = ax \tag{7}$$

where a is the constant of proportionality.

In an infinitessimal time interval, dt, the changes in momentum and displacement coordinates are given by

$$dp_x = ax\,dt$$

$$dx = \frac{p_x}{m}dt \tag{8}$$

Again, because of the linear dependence of dp_x on x and of dx on p_x, the transformation of an initial rectangular action diagram will be into a parallelogram of the same area (see Figure 5). Furthermore, subsequent

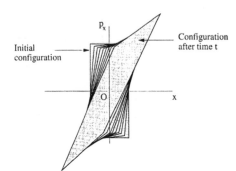

FIGURE 5
Action diagrams resulting from the application of a force in the x-direction on a collection of identical particles when the force is proportional to the x displacement of the particles. The action diagrams are shown overlaid for a sequence of short time intervals.

transformations at a later time will transform parallelograms into new parallelograms of the same area. (This transformation mode can be proven by noting the detailed geometry of the transformation or by using linear algebra, but will not be repeated here.) After a finite time, the initial rectangle will be transformed into a parallelogram. The difference between this transformation and those of a simple drift situation is that here all sides of the parallelogram are rotated.

Before discussing some important special cases of this sort of transformation, it will noted that Equation 7 is the highest order of force field that results in such "linear" transformations. Transformations resulting from forces that depend on higher powers of the displacement coordinates will distort straight lines in an action diagram into curves. (In addition, in real systems where such higher order force fields necessarily involve force dependence on perpendicular coordinates, the area of the action diagrams will not be preserved.) The constancy of the area is an important feature in the use of action diagrams for diagnosing nonlinear distortions, or aberrations, in a charged particle manipulation system.

1. Beam Focusing and Defocusing

An important case of a force acting on a collection of particles in proportion to their displacements is the application of a very strong force for a duration which is so short that movement of the particles is negligible during its application. This case is particularly important in beam optics, where such a force results in a "thin-lens" focusing or defocusing of a beam.

The form of Equation 8 for this special case is

$$\Delta p_x = (a\,dt)x$$
$$\Delta x = 0 \tag{9}$$

When the coefficient in the proportionality between Δp_x and x is positive, the force is a defocusing force; when it is negative, the force is a focusing force. The resulting transformations on a rectangular action diagram are shown in Figure 6.

The action diagrams for the use of a focusing lens to refocus a beam diverging from a previous focus are shown in Figure 7. These action diagrams show that a focus to a narrower width can be achieved, but only at the expense of a higher divergence. They show also that the product $\Delta p_x \Delta x$ is conserved.

It should be noted also that the action diagrams of Figure 7 apply to longitudinal as well as transverse momentum changes. Thus, a force that changes the axial momentum in proportion to the negative displacement of a particle results in a time focusing of a beam pulse. The simplest device with which such a force may be realized is a narrow gap with an accelerating potential which increases with time. The earlier particles, therefore, receive a lower energy gain than do the later particles. The energy-time action diagrams for such a case are shown in Figure 8.

It can be seen once more that the action area is preserved in the refocusing operation, which implies that the energy spread of a beam pulse can be changed, but only at the expense of its duration. Thus, a beam with an axial emittance of, e.g., 6 eV-μs can be focused to a pulse of 1 μs with an energy spread of 6 eV, or to a pulse with an energy spread of 1 eV but with a duration of 6 μs. Therefore, the energy spread of a beam

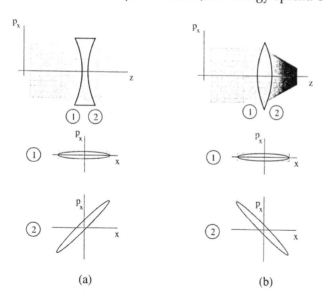

(a)   (b)

FIGURE 6

Action diagrams for (a) defocusing and (b) focusing elements in a beam transport system. Side views of the beam are shown at top. The action diagrams before the operation of the elements are labeled as 1, those immediately after the operation as 2. The ellipses shown are those that fit exactly within the rectangle and the parallelograms.

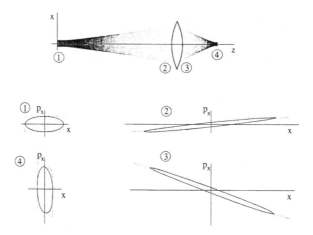

FIGURE 7

The top figure shows the x-z projection of a beam in the z-direction after diverging from a previous focus. Below it, and labeled 1 to 4, are the p_x-x action diagrams for the points labeled 1 to 4 on the x-z projection. The ellipses shown are the largest that can be enclosed by the rectangles and the parallelograms of the action diagrams.

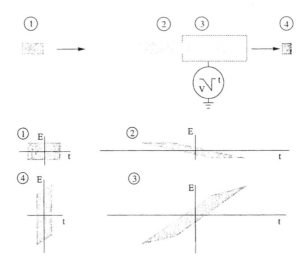

FIGURE 8

Energy–time action diagrams for a beam pulse which is time-refocused after diverging from a previous time focus (as for Figure 7). The refocusing is achieved by having the beam pulse enter a cavity which is being ramped downwards in potential as the beam enters it. The later-arriving particles are given, therefore, more kinetic energy. This extra energy is preserved by returning the cavity potential to zero before any particles leave it, thus allowing the previously slower particles to catch up with the previously faster particles. For clarity, the horizontal scale is exaggerated in the action diagrams.

pulse is often not as useful a measure of the beam pulse quality as is the area of its longitudinal action diagram. In pulse beam manipulation, this area is usually referred to as its longitudinal emittance.

2. Particles Undergoing Simple Harmonic Motion

The simplest form of trapping force is that in which a particle experiences a negative force which is proportional to its displacement. The phase space dynamics of particles experiencing such a trapping force are of particular importance in the trapping of charged particles. The motion that results from the long-time application of such a steady force is, of course, simple harmonic. The trajectory of a particle undergoing simple harmonic motion in an action diagram is an ellipse centered on the origin and with its axis aligned with the momentum and displacement axis (Figure 9).

When a collection of particles of the same mass experience the same force field, they oscillate at the same frequency, hence, in the action diagram of Figure 9b they all move in circular trajectories with the same angular velocity. However, they can have different amplitudes and phases of oscillation. When the particles are distributed randomly in phase and amplitude, then at any particular time the collection randomly fills the circle of the action trajectory for the particle in the collection having max-

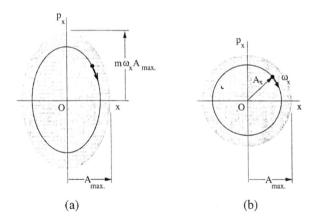

(a)   (b)

FIGURE 9

Trajectories in the p_x-x action diagram of identical particles undergoing simple harmonic motions in the x coordinate at different amplitudes and phases. (a) An arbitrary scaling of the momentum and displacement coordinates. The points for all the particles orbit clockwise in ellipses of the same ellipticity and orbit frequency. A_{max} is the oscillation amplitude of the particle with the greatest energy. The maximum momentum of this particle is mxA_{max}. Particles of lesser energy orbit in smaller ellipses. (b) A scaling of the momentum coordinates that results in the ellipses being circles. In this diagram, all the particle points will move clockwise on circles at a uniform angular velocity equal to the oscillation angular frequency.

imum energy. This circle is, therefore, the permanent shape of the action diagram for such a collection. In time the individual particles move counterclockwise within this ellipse. It is clear that the phase space volume of this collection is preserved.

Elliptical action diagrams are not only of importance in describing collections of particles undergoing simple harmonic motion; they also describe the typical collections of particles in beams because of the circular aperture of most beam-delivery systems. Thus, while an emittance diagram which is initially a parallelogram is useful for determining aberrations in a system, the typical beam emittance diagram is an ellipse. For this reason, ellipses are often shown within parallelogram action diagrams, as in Figures 6 to 8, and the emittance of a beam is usually given in units of π-mm-mrad.

Because the area of an ellipse is π times the product of the semimajor and semiminor axes, the emittance of an elliptical diagram, when expressed in these units, is simply the product of the maximum divergence in milliradians and the maximum displacement in millimeters at a focus. For reference, it is convenient to note that the transverse action of a 10-keV beam of particles of $m = 100$ Da with emittance 1 π-mm-mrad is 0.45 eV-μs. A typical emittance of such a beam from a high-quality ion source would be about 5 π-mm-mrad corresponding to a transverse action of about 2 eV-μs.

Another significant point of elliptical action diagrams is that in linear systems, elliptical action diagrams remain elliptical, just as parallelogram action diagrams remain as parallelograms.

D. Transfer of a Single Beam Pulse into a Trap

The phase space description by action diagrams of the required transformation, of an external beam pulse into a stable collection within an ion trap, is simple. The elliptical action diagram of the beam pulse is to be transformed into the ellipse corresponding to stable motion of the particles within the trap.

1. Axial Phase Space Considerations of Trap Injection

The most obvious requirement for injection of particles into a trap is to have a decelerating electric field which will remove the central momentum of the beam pulse. The simplest such field is uniform in the axial direction. When the deceleration of the central particle of initial energy, E_0, to rest in a distance, z_0, occurs in a time that is long compared to the duration of the beam pulse, but in a time which is sufficiently short that the momentum spread introduced by the deceleration is much larger than the momentum spread of the pulse, the equations relating the coordinates of the original action diagram (E,t relative to the central particle) and the final action diagram (p_z, z relative to the central particle) become simple. They are

$$p_z = \frac{E_0}{z_0} t$$

$$z = \frac{z_0}{E_0} E \tag{10}$$

The action diagrams for a beam pulse which is at a time focus when the deceleration is started are shown in Figure 10.

Thus, the momentum spread of the decelerated pulse is related directly to its initial time spread, because the reduction in momentum of a particle in a decelerating field is proportional to the time the particle is in the field. Similarly, the spatial spread of the decelerated pulse is related directly to its initial energy spread, as the distance that the particle travels in this field in coming to rest is proportional to its original energy.

Of course, as shown in Figure 10, a beam pulse which comprises an action ellipse will transform into an action ellipse at rest. Thus, if one desires to have such a focused beam pulse of energy spread ΔE and duration Δt adapted to the ellipse for simple harmonic motion within a trap, the ratio of the incoming central energy of the particle (E_0) and the deceleration length (z_0) should be chosen in order to give the correct rela-

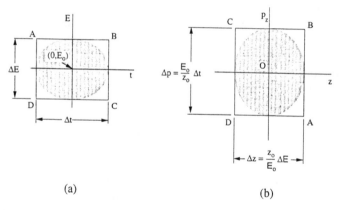

(a)   (b)

FIGURE 10

Action diagrams for a beam pulse that enters a uniform deceleration field with a central energy E_0 and focused in time (a), and that is decelerated in this field for a time which brings the central particle exactly to rest (b) in a distance z_0. These diagrams are valid for the assumptions that the initial energy spread is very small compared to the central energy and the time spread is small compared with the time for deceleration. The shaded area representing the particle collection is shown enclosed in the rectangles specified by the action diagram limits.

tionship between the momentum and the spatial spread for such motion. As indicated by Figure 9, this condition leads to

$$\Delta p_z = m\omega \, \Delta z \qquad (11)$$

where m is particle mass and ω is the angular frequency of the simple harmonic motion within the trap.

From the values for Δp_z and Δz shown in Figure 10, and the relationship given in Equation 11, the ratio of E_0 to z_0 is given by

$$\left(\frac{E_0}{z_0}\right)^2 = m\omega \frac{\Delta E}{\Delta t} \qquad (12)$$

The most convenient method for decelerating ions into a trap is to use the electrodes of the trap itself for the deceleration. By bringing the far end-cap electrode up to a potential equal to twice the central energy of the incoming beam pulse and the ring electrode up to a potential just equal to that energy, a fairly uniform decelerating field is produced which will bring the ions to zero central velocity at the center of the trap. Thus, the deceleration length z_0 is just half the separation of the trap end-cap electrodes.

In this way, the incoming energy E_0 is determined. An incoming beam pulse of a different energy would have to be either accelerated or decelerated to this energy before approaching the trap itself. Alternately, the beam pulse could be time-refocused to obtain the correct relationship for Equation 12.

As a representative example, suppose that a trap with 20 mm between the end-cap electrodes forms a simple harmonic potential well which is 100 eV deep for ions of $m = 100$ Da. Ions with an amplitude of oscillation of 10 mm would then have an energy of 100 eV or a momentum of 14.4 eV-μs/mm. Thus, the proportionality constant, $m\omega$, between Δp_z and Δz is 1.44 eV-μs/mm². If the beam pulse to be captured has an energy spread of 1 eV and a duration of 0.1 μs, then Equation 12 indicates that the incoming energy of the beam should be adjusted to 38 eV. The deceleration time for the central energy will be 1.17 μs. A pulse of 76 V must be applied, therefore, to the far end-cap electrode for up to a time of 1.17 μs after the arrival of the central particle at the trap.

The resulting momentum and axial spreads of the decelerated pulse will then be 0.38 eV-μs/mm and 0.26 mm, respectively, which compares with a momentum spread in the beam pulse of 72 meV-μs/mm centered on 14.4 eV-μs/mm, and a beam pulse length of 1.39 mm. These figures show that the assumptions leading to Equation 10 are valid for this example.

The longitudinal emittance of the beam pulse in the previous example, i.e., 0.1 eV-μs, is representative of that which can be achieved by pulsed laser ionization of atoms or molecules. It is seen that the beam pulse fills very little of the available phase space of the trap ($\pi \times 14.4$ eV-μs/mm $\times$ 10 mm $\approx$ 450 eV-μs). Thus, unless the ion pulse is to be localized precisely on the trap center, there is a great deal of leeway for error in the various parameters of the capture system. Theoretically, with precise manipulation, up to 450 μs of a beam of 1 eV spread could be accepted by the trap, which is about 1500 times as much as that given in the example. However, such adaptation could not be achieved by the simple application of a steady decelerating field for a specific time. Rather, both the energy of the incoming beam and the retardation field would have to be tailored carefully to vary with time until the trap was full. Such tailoring is possible with modern programmable waveform generators, but the proper programming of such generators would require accurate detailed knowledge of the longitudinal emittance of the incoming beam.

2. Transverse Phase Space Considerations of Trap Injection

The transverse phase space considerations of trap injection are somewhat simpler than the axial considerations because the transverse action diagrams of a beam show directly the momentum-displacement relationships of the particles. All that is required in order to focus the beam so that the action diagram of the focus is just that which the particles would have in the trap is that the strength of the focusing be adjusted at injection to give the correct ratio between the momentum spread and the displacement spread of the beam. As an example, suppose the beam pulse to be trapped had a transverse action of 2 eV-μs, corresponding to an emittance of about 5 π-mm-mrad for 10 keV ions of $m = 100$ Da. If the transverse simple harmonic motion in the trap has the same frequency as for the axial motion in the previous example, then the ratio of the momentum spread to the displacement spread should be 1.44 eV-μs/mm^2. This condition requires that the focus at injection should have an action diagram which is about 1.2 mm wide and 1.7 eV-μs/mm high. Ideally, the decelerating system to take the beam pulse down to the selected injection energy (38 eV) should be designed to achieve a transverse focus of this strength. For an injection energy of 38 eV and $m = 100$ Da, a transverse momentum of ± 0.85 eV-μs/mm corresponds to a divergence of about ± 100 mrad.

While giving perhaps some indication of the scale of decelerating and focusing strengths required for precise injection into the center of a trap, the above examples should not be regarded as precise solutions for such injection. Because the onset of the retardation electric field of the trap necessarily involves derivatives of that field which, from the considerations of Section II.C, must in themselves give focusing forces, the transformations of the action diagrams that an incoming beam pulse will undergo can be predicted accurately only with detailed numerical integration of the particle motions. The above considerations should give an indication of the range of system parameters to investigate in any such detailed calculations.

In any case, it is seen that there is wide latitude in both the axial and transverse dimensions for injecting a single beam pulse into a trap. A simple system for efficient collection of a pulse of laser desorbed ions into a Paul trap has been presented by Davey and co-workers.[6]

However, many cases exist in which the available material cannot be efficiently ionized as a pulse. One such example is the ionization of radionuclides produced in thick targets by high-energy proton bombardment at the ISOLDE facility at CERN (Geneva). There, the radionuclides are boiled out of the target material over a period of up to several seconds. The engineering of systems for collecting a beam of this duration into a trap is, therefore, of interest.

III. CONTINUOUS BEAM INJECTION INTO TRAPS

A. General Phase Space Considerations of Continuous Trap Injection

The design of injection systems for the collection of a continuous beam into an electromagnetic trap poses several challenges. The first is that the axial action space of the trap is limited. For the example taken in Section II.D.1 of a trap with 20 mm axial space and a well depth of 100 V for m = 100 Da, the axial action space is about 450 eV-µs. A more typical trap of these dimensions would have a well depth of about 30 V, for which the axial action space would be about 250 eV-µs.

A typical direct current (DC) beam that one might wish to collect in a trap would have an energy spread of about 5 eV. A typical electromagnetic trap of 250 eV-µs axial action space could, therefore, accept only about 50 µs of such a beam using straightforward axial deceleration of the beam, and even this could be achieved only by careful programming of the injection energy into the trap and of the retardation voltages.

However, a great deal of phase space is available in the transverse coordinates of a typical trap. A typical beam from a radionuclide separation facility will have an emittance of about 2 π-mm-mrad at 60 keV, corresponding to a transverse action of about 2.2 eV-µs. A transverse action space of 250 eV-µs in a trap would be over 100 times greater than this. In theory, therefore, 100 beam pulses of axial action 250 eV-µs could be stacked side by side in the phase space of each transverse coordinate.

Because there are two transverse coordinates, up to 10^4 beam pulses could be collected theoretically in a trap before the total phase space capacity of the trap is exceeded. Theoretically, if each beam pulse is of 50 μs duration, then a total of 0.5 s of beam could be fitted into the trap.

Multiple stacking of successive beam pulses into different regions of transverse action space is a common technique used for loading a synchrotron ring from a linear accelerator for subsequent acceleration in the ring to very high energies. The number of successive pulses that can be stacked is typically 100. Taking this value as a more realistic number for the successive pulses that can be stacked into an electromagnetic trap, it should be possible to collect about 5 ms of a typical beam into a typical trap.

Stacking more than about 5 ms of a typical beam into a typical trap will require condensation of the phase space of the particle collection already within the trap. Fortunately, such condensation is achieved relatively easily in electromagnetic traps by buffer gas cooling. Times of several milliseconds have been observed for the cooling of ions to several electronvolts in a Paul trap operated in helium at 0.01 Pa (10^{-4} mbar).[2] Under such cooling, the phase space volume of 5 ms of beam collected in a trap would have shrunk considerably by the time the collection is completed. Thus, more beam can be accepted and, if the cooling rate is sufficiently high, beam may be added to the trap almost indefinitely.

However, there are several practical limits as to how long an electromagnetic trap with cooling can collect a DC beam. One such limit is the space-charge limit of the trap. A typical Paul trap will not hold $>10^6$ ions before the coulomb repulsion between the ions forces them onto the electrodes of the trap. The useful collecting time for the trap is the time it takes for the beam to deliver 10^6 ions to the trap.

Another limit on the useful collection time is the lifetime of ions in the trap, where the lifetime depends on the purity of the helium in the trap and on the chemical nature of the ion. Using high purity helium at 0.01 Pa, lifetimes of >1n have been observed for cesium ions,[7] but in a similar trap using similar purity of helium, the lifetime observed for xenon ions appears to be on the order of tens of seconds.[8]

B. Practical Systems for Continuous Trap Injection

While general phase space considerations lead to the theoretical possibility of continuous collection of a DC beam in a trap until the space-charge limit is reached, practical beam manipulations to achieve this are not straightforward. The principal difficulty is that the oscillation frequency for axial motion within the trap is typically 100 kHz. Ions that arrive at the trap with practically zero energy and which then fall into the trap will, even if they do not fall onto an electrode in their first oscillation, still return to the injection point within 10 μs unless sufficient axial energy is removed or unless sufficient transverse motion has been induced to prevent this from happening. For single-pulse injection, discussed in Section II.D, the removal of axial energy can be achieved by pulsing the trap electrodes. However, for multiple-pulse injection, such pulsing of the trap electrodes will drive out ions that are already in the trap.

The problem presented here is how to disturb the axial and transverse motion of the incoming ions to prevent their immediate escape without overly disturbing the ions that are already in the trap from previous cycles. Two approaches have been taken toward solving this problem. One approach is to apply an axial electric field oscillating at a frequency close to that of the axial oscillations and at a phase that dampens the axial oscillation that follows injection of a low energy beam pulse into the trap. Such was the approach taken by Moore and Gullick,[7] who found that about 50 percent of incoming pulses of about 2 eV could be captured in a Paul trap until the space-charge limit of the trap was reached.

A second approach is to transform some of the axial energy into transverse energy so as to reduce the amplitude of the subsequent axial oscillations. This transformation must be done by deliberately introducing higher order aberrations in the fields acting on the ions entering the trap. This approach was taken by Rouleau[8] when injecting higher energy ions (about 20 eV) into a typical Paul trap at the ISOLDE facility at CERN. In that work, a collection efficiency of 0.2 percent of a 60-keV DC beam of xenon ions was achieved.

Another possible approach that has not yet been tried is to treat the problem of trap injection in a manner similar to the injection of beams into cyclotrons. Here, the beam is bunched into the RF cycles of the cyclotron and phased at injection so that the incoming bunches are accelerated quickly away from the center of the machine. In this way, the transverse phase space of the machine is filled continuously. A similar effect could be produced in a Paul trap by splitting the end-cap electrodes into quadrants and applying an azimuthal quadrupole field at the transverse oscillation frequency of the trapped ions. A rotating electric field would be produced that would induce the ions to move in expanding circles around the trap axis, thereby moving them away from the injection hole of the trap and into a region where, in a trap with hyperbolic electrodes, there is greater freedom for the axial motion. Of course, such a rotating azimuthal quadrupole field should not be so strong as to move the ions to the ring electrode.

The most direct approach to achieve high collection efficiency in a trap is to make the trap large and deep enough so that the small phase space volume of the incoming beam pulses has but a small chance of returning to the injection region of this volume within the cooling time. This approach has been taken in the design of a practical system for collecting ISOLDE beams at the new facility at the CERN booster accelerator. The design is based on the results of the preliminary investigations of collection efficiency,[8] in which a rapid rise in trapping efficiency was observed for increased phase space volume of the trap (Figure 11). The new trap is to have an end-cap electrode separation which is three times that of the one with which the results shown in Figure 11 were obtained, and is to be operated at up to 10 $kV_{(0-p)}$ RF amplitude. From an extrapolation of the results shown in Figure 11, it is expected that the trap should reach total capture of ions appearing within the RF window of acceptance of the trap (see below).

C. Effects of the Radiofrequency Phase on Paul Trap Injection

In any collection of an ion bunch in a Paul trap the effects of the phase of the electric field of the trap itself must be taken into account. When the potential on the ring electrode is at a maximum, the field will be highly repulsive for positive ions entering the trap through an end-cap electrode, repelling low energy incoming ions back out of the trap. When the potential on the ring is at minimum such ions will be pulled into the trap, gaining considerable energy from the field. Thus, the subsequent stability of the ion motion and, consequently, its containment within the trap, is greatly dependent on the RF phase at which it enters.

This effect has been discussed by Todd et al.,[5] who calculated the properties of ion beams injected into a trap with the quadrupole field turned off and then turned on at a specific time after injection. The effect was also clearly observed in the work of Moore and Gullick,[7] in which different phasings of the incoming beam pulses relative to that of the RF

FIGURE 11
The efficiency of collection of a DC beam of xenon ions from the ISOLDE facility into a Paul trap as a function of the RF voltage amplitude applied between the ring and end-cap electrodes of the trap. The trap had electrodes which were approximately hyperboloids of revolution and the end-cap electrode separation was 25 mm. The dashed curve on the diagram corresponds to a rise of efficiency in proportion to the cube power of the RF amplitude.

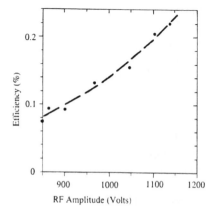

drive potential of the trap could change the collection efficiency from the full 50 percent to zero.

The phase window in which specific energy ions can be injected into a Paul trap for subsequent stable capture is very difficult to determine. Rough calculations[7] indicate that the phase window is dependent on the injection energy and for energies up to about the pseudopotential well depth of the trap, may be as high as 10 percent of an RF cycle.

If the RF window for collection of a beam into a Paul trap is indeed 10%, the collection efficiency within this window reported in the work of Rouleau[8] was about 2%. In other words, if the DC beam from the ISOLDE facility could have been bunched into the RF windows of the Paul trap, the collection efficiency would have been raised from 0.2 to 2 percent.

D. Design of Deceleration Systems for Trap Injection

The collection of a high velocity beam such as the 60-keV beams of the ISOLDE facility requires very careful design of the decelerating system in order to achieve the phase space requirements for successful collection in an electromagnetic trap. However, it was found in the work of Rouleau[8] that deceleration to axial energies of no more than 20 eV was necessary. To achieve this degree of deceleration, the transfer of axial energy into transverse energy during the deceleration had to be kept to a minimum. In phase space terms, the deceleration system should not couple axial phase space components to the transverse components. In practical terms higher order multipole components beyond the quadrupole in the decelerating electric field must be made insignificant. Yet higher order multipoles in an electric field which operates over a short distance are unavoidable. The aim for an accelerator design is that the higher order multipoles be weak enough not to have a significant effect on the action diagrams.

The decelerator design that produced the correct focus for injecting beam into the Paul trap in the work of Rouleau[8] without significant aberrations is shown in Figure 12. This design was based on electric field calculations by finite difference relaxation of the potentials at grid points followed by accurate simulation of ion trajectories through these calculated fields using sixth-order Runge-Kutta integration with adaptive stepsize control by error monitoring. To achieve the necessary accuracy (about 1 eV for a deceleration of 60 keV), the electric fields at the trajectory points were interpolated from the grid point potentials by fitting electric multipoles up to sixth order (dodecapole) to these potentials and calculating the interpolated fields from these multipoles. The principle of this method is given by Lunney and Moore.[9]

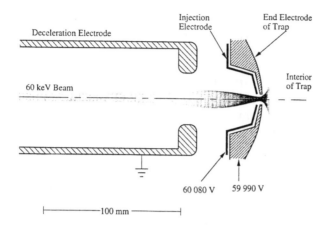

FIGURE 12

The deceleration system used for preparing a 60-keV ISOLDE beam for injection into a Paul trap. The transverse extent of the beam is exaggerated by a factor of 5 for clarity. The injection electrode is set at a potential such that the potential on the axis reaches about 59,980 V at the trap entrance.

IV. PHASE SPACE DISTRIBUTIONS WITHIN A TRAP

Immediately after collection of an ion pulse in a trap, the phase space distribution of the ions will be that of the original beam pulse modified by the electric fields of the injection system and of the trap itself. After some time, ion/ion interactions and, possibly, ion/molecule interactions in the face of significant background gas, will randomize the phase space distribution. The resulting distribution will be that which is most probable statistically. This section describes typical distributions for electromagnetic traps.

A. Phase Space Distributions in an Ideal Trap

The most probable distribution of a total energy E among N particles is the Gibbs distribution

$$\frac{dN(E)}{dE} = \frac{N}{\bar{E}} e^{-\frac{E}{\bar{E}}}$$

(13)

where $dN(E)$ is the number of particles with energy between E and E plus dE and $\bar{E}$ is the average energy of the particles, i.e., E/N. The fundamental assumption of statistical mechanics is that Equation 13 is, therefore, the distribution of energies that will actually result after a large number of interactions between a large number of particles when the interactions randomly distribute a fixed total amount of energy.

Another important result of statistical mechanics is that the total energy will be divided equally between the canonical coordinates of the motion. The total energy in each of the canonical coordinates is therefore constant, and Equation 13 applies to the distribution of energy in any one of these canonical coordinates. Furthermore, the average energy in each of these coordinates is the same and becomes a characteristic of the particle collection. In many cases, it is appropriate to define a temperature T for the collection in terms of this average energy as

$$\overline{E}_r = kT \tag{14}$$

where x is a representative canonical coordinate and k is Boltzmann's constant.

In an ideal trap with independent simple harmonic motions in all three Cartesian coordinates, the canonical coordinates are these Cartesian coordinates. The Gibbs distribution (Equation 13) applies, therefore, to the action diagram for each coordinate. For simplicity, consider an action diagram in which the momentum is scaled so that the particle trajectories for simple harmonic motion are circles, as in Figure 9b. The energy E of a particle of amplitude of oscillation a is

$$E = \frac{1}{2}m(\omega a)^2 \tag{15}$$

The energy spread in an annular band of width da is, therefore,

$$dE = m\omega^2 a \, da \tag{16}$$

and from Equation 13 and 14 the number of particles in this band is

$$dN(E) = \frac{N}{kT} e^{-\frac{E}{kT}} m\omega^2 a \, da \tag{17}$$

In a random distribution, the particles in this band will be distributed uniformly in phase of oscillation, or uniformly spread around the band. Using Equation 15 for E and noting that the area of the annular ring dA is $2\pi a \, da$, the density per unit area of the action diagram is seen to be the Gaussian

$$\frac{dN(E)}{da} = \frac{N}{2\pi kT} m\omega^2 e^{\frac{-m\omega^2 a^2}{2kT}} \tag{18}$$

Integrating this distribution over the momentum variable p_x in a displacement slice dx will give the distribution in the spatial coordinate x. This distribution is easily seen also to be a Gaussian of the same half-width as that given by Equation 18.

Although the action distribution for a Paul trap is severely distorted from that for an ideal simple harmonic oscillator trap (see Section IV.C), Gaussian spatial distributions have been observed for such traps,[10] confirming the basic statistical nature for the distributions.

B. Effects of Space Charge in an Ideal Trap

The elementary statistical model of the action diagrams for particles in the ideal trap described above assumes no significant effects from the Coulomb force of the collection on an individual particle. When this effect is significant, it will modify those diagrams. Dehmelt[11,12] showed that in the case of an ideal trap forming a spherical potential well (i.e., the simple harmonic frequencies are the same for all three coordinates), and where the statistical motion is insignificant (i.e., the particles are cold), the particles will form a sphere of uniform density and of radius such that the Coulomb force, for particles at the edge of the sphere, balances the restraining force of the potential well. None of the particles have any momentum (the Coulomb forces balance the trapping electric field throughout), therefore, the action diagrams are just horizontal lines centered on the origin. Such lines, of course, have no action area, so even though the particles occupy a physical volume, their phase volume is zero, essentially because they have no momentum.

Ion collections in which the space charge and the temperature effects are comparable would have action diagrams intermediate between the ellipses of Figure 9 and horizontal lines centered on the origin. These diagrams can be thought of as the elliptical diagrams of the pure simple harmonic oscillation stretched in the horizontal direction by the space-charge forces. Because the Coulomb force on a particle is strictly a function of its position (it cannot have a statistical variation at a particular position), it does not contribute to the phase space volume of the particles. The stretching of the simple harmonic motion action ellipses by the Coulomb force preserves their action area. However, because the Coulomb force is not a linear function of coordinate displacement, the stretched shapes that result are not actually ellipses.

C. Phase Space Distributions in a Paul Trap

The ideal ion trap, i.e., the three-dimensional simple harmonic oscillator, cannot be created by manmade electromagnetic fields. The closest that can be achieved is the Paul trap in which there is indeed simple harmonic motion in all three Cartesian coordinates (commonly called the "macromotion"), but upon which is superimposed an oscillation that is in phase with the RF drive potential of the trap (commonly called the RF motion).

The action diagram for a collection of particles in a Paul trap at a particular RF phase is important when considering injection of an incoming beam bunch. For optimal injection, the retardation system must be designed to bring the central momentum of the particle bunch to rest; the action diagram of the particle bunch will then conform to that for equilibrium within the trap at the RF phase at that instant. The action diagrams are particularly important when considering extraction of the collection for observation by other instruments. For example, when the collection is to be mass analyzed by a TOF mass spectrometer, then the extracted bunch must be time-focused at the TOF detector. To achieve this objective, and to ascertain the mass resolution that can be obtained with a given mass spectrometer concept, one needs to know the action diagram of the particles in the trap at the moment the extraction operation is started.

The macromotion in a Paul trap can have any phase and amplitude and, therefore, in a statistical equilibrium will have an action diagram similar to that for the ideal simple harmonic motion trap. The distribution of the particles within this action diagram can be characterized, therefore, by a temperature just as for the action diagram of the hypothetical ideal trap discussed in Section IV.A.

However, because of the RF motion superimposed on the macromotion, the action diagrams for particles in a Paul trap will not be ellipses with axes which are aligned permanently with the momentum-displacement axis as shown in Figure 9. To a first-order approximation, the RF motion is characterized by having a displacement in phase with the RF voltage and an amplitude that is proportional to the electric field at the center of the motion. Because the electric field is proportional to the distance from the trap center, the amplitude of the RF motion will be proportional in turn to this distance. Its effects on the action diagrams can be visualized by considering an action line for a collection of particles, all of which have the same macromotion phase (Figure 13). Because all the particles are in phase with the RF, they will all be on ellipses centered on this line, and all with the same azimuth. Therefore, all will be on a line which is rotated with respect to the macromotion line.

The same will be true for lines representing all other phases of macromotion. The overall result on the full action diagram for the macromotion will be as shown in Figure 14. These pictures were obtained by numerical simulation of a collection of ions in a representative Paul trap. Similar pictures are presented elsewhere.[4,13,14]

Figure 14 demonstrates the usefulness of action diagrams in understanding the motion of particles in traps. In the macromotion action diagram, the particles simply orbit clockwise in circles at the macromotion angular frequency. The RF motion, superimposed on this motion to give the actual motion of the particles, simply distorts this circle into an el-

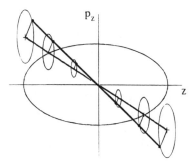

FIGURE 13

RF distortion of the action line for a collection of particles that are all of the same phase of macromotion. The straight line representing the macromotion of the particles is shown shaded. The large ellipse is the macromotion trajectory for a representative particle. The smaller ellipses are for the RF motion of a selection of six particles. The phase of this motion is indicated by the small lines within these ellipses. The straight solid line is the actual action line for the particle collection at the instant selected.

liptical shape which rotates at the RF angular frequency. Thus, in one RF cycle, when the shape of an action diagram repeats, the particles within this shape will all move clockwise by an amount equal to the progression of the macromotion oscillations.

An important point concerning the RF distortion of the action diagram in a Paul trap is that it preserves the action area. This preservation of the action area may be suspected from the effects of the RF distortion on the macromotion line in Figure 13, but it can be proven by taking into account that the RF distortion is due only to a time variation of a quadrupole field. Because a quadrupole field gives linear equations of motion, it can only distort elliptical action diagrams into ellipses of the same area. Another way of considering this phenomenon is to note that the RF motion is coherent with the RF field. Therefore, it cannot have an independent phase relationship and so cannot contribute to the phase space volume of a particle collection. This point is often misunderstood in dealing

FIGURE 14

RF distortion of the action diagram of particles in a typical Paul trap. (a) The actual action diagram at zero RF phase (zero voltage, rising) superimposed on the action diagram for the macromotion with the momentum axes scaled so that the diagram is a circle. (b) Action diagram for each quadrant of an RF cycle; the diagram for 360° phase is the same as that for zero phase.

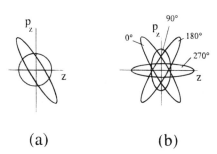

(a)   (b)

with particle motion in Paul traps as the RF motion does contribute significantly to the average kinetic energy of the trapped particles, usually more than does the macromotion. That it does not contribute to the phase space volume and, therefore, to the statistically defined temperature of the trapped particle collection is sometimes confusing. Perhaps the confusion can be dispelled by noting that if the particles were extracted from the trap, the RF distortion of the action diagram would be removed and the action area would be that of the macromotion alone. One must be very careful, therefore, to distinguish between temperature and average kinetic energy of particles in a Paul trap.

As in the discussion of the ideal trap in Section IV.A, the action diagrams for the Paul trap shown in this section are for particles with no mutual interactions. If, for instance, space charge effects are included, then the elliptical action diagrams are elongated in the spatial coordinates in a fashion similar to that discussed in Section IV.B for the ideal trap. In the Dehmelt model[11,12] referred to above, cold particles in a Paul trap fill a volume in a pseudopotential well with uniform density out to a radius at which the force at the wall of the pseudopotential well balances the space charge repulsion. The action diagram for such a collection is simply, as pointed out above for an ideal trap, a horizontal line. However, because the actual electrical potential is much greater than the pseudopotential, there will be RF motion on the particles, and thus, the action line will wobble about the horizontal axis (Figure 15).

In a typical trap, the particles will not be cold, nor will the space charge forces be insignificant. The action diagrams will be intermediate between those of Figures 14 and 15.

Unfortunately, while the general shape of the action diagram of particles in a Paul trap, as described above, are understood, its dimensions and how they vary with type and quantity of particles are not. Essentially what is needed is the temperature of the statistically distributed macromotions that gives the overall size of the action diagram (space charge effects are relatively easily calculated). This temperature is a result of a complicated set of interactions reaching equilibrium. Even though the RF

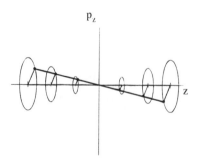

FIGURE 15

RF disturbance of the action line for a collection of cold particles in a Paul trap. The ellipses represent the trajectories for the RF motion of a selection of six particles. The horizontal line on the z-axis represents the cold particles in the absence of RF motion. The solid line represents the action line at the particular RF phase chosen for the diagram.

motion itself contains no energy that can be considered thermal, it is this motion that leads indirectly to the thermal distribution of the macromotions. In detail, the cause is the scattering of particles from truly coherent RF motion by collisions with background gas molecules, or perturbations caused by higher order electric field components within the trap. In a macroview, the cause is friction of the particles against the background gas or the trap perturbations or both. This friction heats the particle motions, a process called RF heating. However, at the same time, the background gas and (possibly) electrodynamic interactions with external circuitry connected to the trap electrodes cool the particle motion. It is the balance of the heating and cooling processes that determine the equilibrium temperature of the macromotion.

The situation is complicated further by the fact that even though the space charge forces also do not contribute directly to the phase space volume of the ion cloud (see above), they are responsible indirectly for the RF heating. This is because particles that are allowed to reach the center of the trap experience no electric field and, thus, no RF motion. (It is assumed here that the trap is electrically symmetrical and, therefore, that the field is zero at the center of the pseudopotential well of the trap. For this to be true, no odd-order electric multipoles can occur and the trap must be electrically symmetrical.) For sustained RF heating, particles must be prevented from reaching the trap center (they are, therefore experiencing RF motion). In a typical Paul trap, it is the space charge forces that do this.

Because of the uncertainties concerning all of these interactions, even modeling the thermodynamics within a Paul trap on a powerful computer is of little value. The most useful results are those of actual measurements of spatial and momentum distributions in a trap. Unfortunately, such measurements are very difficult to carry out and some give inconsistent pictures of the phase space volume. A literature review of such measurements is given by Lunney et al.[14]

V. EXTRACTION OF PARTICLES FROM A TRAP

As mentioned in Section I, the possibility of preparing a collection of ions for introduction into a specialized instrument is one of the most attractive features of electromagnetic traps. If the ions can be cooled even to room temperature in a trap, their phase space volumes will be much smaller than that of typical ion sources in which temperatures are measured in electronvolts (1 eV is about 40 times room temperature). Modern laser cooling methods have cooled single ions in Paul traps to the range of milliKelvins.

Unfortunately, because of space charge repulsion, electromagnetic traps can hold only a limited number of charged particles. For a typical trap, the limit is about 10^6 singly charged ions with $m = 100$ Da. Traps are most useful for dealing with trace quantities of material where their high sensitivity can be a very great advantage.

To make the maximum use of a collection of trapped particles in an external instrument, the collected ions must be extracted efficiently and with as little disturbance as possible to their phase space volume. In general, two classes of extraction are of interest: slow extraction into an almost continuous beam and fast extraction as a sharply defined pulse. This situation is similar to the case of storage rings for high energy particles where different experiments demand either one or the other of these two types of extraction.

A. Slow Extraction from Traps

Slow extraction of collected particles in a trap is similar, in principle, to that used for high energy accelerators. A resonant oscillation is usually set up by a small perturbing field so that the phase space volume of the particles is elongated in the extraction coordinate until it exceeds the physical limits of the device. In the case of an electromagnetic trap for ions, it is usual to excite the axial motion by applying a small tickle voltage to one or both of the end-cap electrodes until the ions leave the trap through a hole in one of the electrodes. After exiting through the hole, the ions can be accelerated to any desired kinetic energy by DC electric fields. The use of such slow extraction as a very sensitive method of detecting stored ions in a Paul trap is described by Vedel et al.[15]

The important considerations in slow extraction are the energy spread, emittance, and duty cycle (roughly defined as the ratio of pulse beam duration to the interval between beam pulses). The energy spread and duty cycle are controlled by the strength of the perturbing field; the lower this field, the lower the energy spread and the higher the duty factor. The emittance of the extracted beam, assuming that the perturbing field does not have aberrations sufficient to cause blowup of the transverse action diagram, is governed by Equation 6, in which A is the transverse action area of the ions in the trap. The energy spread and the duty cycle of the beam in very slow extraction will depend on the action area of the axial motion in the trap and details of the extraction procedure. By gradually lowering the trapping electric field in a Penning trap and using small electric pulses on one of the end-cap electrodes, the system used for the work of Reference 1 can achieve ion extraction over several milliseconds with energy spreads on the order of several millivolts.

The effect on the action areas of slow extraction from a Paul trap with buffer gas cooling are not as predictable as for a Penning trap because of RF heating (see above) during extraction. Paul traps are more attractive, therefore, as devices for producing short duration pulsed beams in which the extraction process occurs sufficiently quickly so as to avoid RF heating.

B. Fast Extraction from Traps

The important considerations in fast extraction from a trap are the transverse and longitudinal emittances of the extracted beam pulse. The extraction of an ion collection from a trap as a beam pulse is the reverse of the process of capturing an incoming beam pulse in the trap. An electric field is applied for a duration required to complete the extraction. In general, the higher the electric field used to extract the collection, the higher the energy spread and the shorter the time spread of the extracted pulse. Again, phase space considerations are useful, even at an elementary level. At a time focus after extraction with linear fields to particle momenta high compared to those within the trap, the product of the particle energy spread and the pulse duration will be a constant, and equal to the action area of the collection in the trap before extraction.

The importance of applying a linear field can be seen in a computer simulation of the action diagram for a collection of particles extracted from a Paul trap by applying a high voltage pulse to one of the end-cap electrodes (Figure 16). The collection shown in this figure started at the trap center as a rectangular array. Because the electric field was obtained by applying a potential to one electrode only, a large sextupole component of the field gave rise to the curvature of the action diagram. This curvature could have been removed by applying a potential to the ring electrode equal in magnitude to half the potential applied to the end-cap, or by applying an equal and opposite potential to the other end-cap electrode. Both of these arrangements would have removed the sextupole component, leaving the octopole component as the most significant higher order multipole. (An octopole component would give an S-shaped curvature to the edges of the action diagram, and no such curvature is noticeable in Figure 16.) An alternate method to eject ions from a trap without high order multipole aberrations would be to apply a very high pulse for a very short time to the extraction end-cap electrode. The principle of this method is to impart the momentum required for the ions to leave the trap in a sufficiently short time that the applied fields act only on those particles that are near the trap center. Because all multipoles of the electric field except the dipole (uniform field) are zero at the trap center, the effect of the higher-order multipoles can be made insignificant. If the

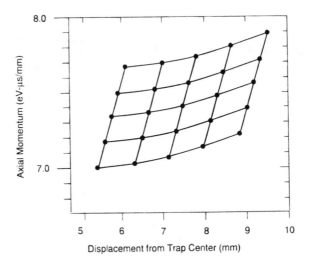

FIGURE 16

Action diagram for a collection of particles that have been displaced near one of the end-cap electrodes of a Paul trap by a high electric potential on that electrode. The action diagram at the center of the trap was a square array for the scales shown on this plot. The curved lines are drawn to guide the eye.

extraction process is indeed linear, then the transverse and longitudinal emittances of the extracted beam pulse can be obtained from the action areas of the stored ion collection.

For a given extraction procedure for a Paul trap, the action diagram of the extracted beam pulse will depend a great deal on the RF phase at which the extraction procedure is initiated, as action diagrams within the trap vary considerably with RF phase (Section IV.C). This problem was first discussed thoroughly in a series of papers by Mather, Todd, and Waldren of the University of Kent.[16-20] Figure 17 shows what can be expected for a representative Paul trap. The action diagrams shown in this figure were obtained by computation of the energies and times of arrival of ^{39}K ions delivered at about 900 eV to a detector approximately 700 mm from the trap center. The calculations were for a trap with a purely quadrupole field (in the absence of the extraction field) and with 28.72 mm separation between the end-cap electrodes. The RF voltage amplitude was 420 $V_{(0-p)}$ and frequency 0.65 MHz, giving a macromotion frequency of 151 kHz. No DC component was used. Extraction was effected by applying a potential of 200 V to an end-cap electrode at particular phases of the RF. The simulations were carried out for an ion collection occupying 2 mm of axial extent in the macromotion. The action diagram for the macromotion of such a collection immediately before extraction is shown in Figure 17a; the individual particles are coded from 1 to 25, as shown.

The only parameter varied for Figure 17b, c, and d was the phase at which the extraction potential was applied. Figure 17b shows the action diagram when the ion collection is extracted at the RF phase of 197°, which gave the minimum TOF to the detector. In Figure 17c, the collection was extracted at an earlier RF phase of 94°. A significant increase occurs in the beam pulse duration which can be seen from the coding of the points, due to the original momentum spread in the macromotion. In Figure 17d, the collection was extracted at a later RF phase of 215°. It is seen here also that a significant increase occurs in beam pulse duration, but is now contributed mainly by the amplitude of the original macromotions.

Figure 17 shows also that neither the energy spread nor the pulse duration is a good indicator of beam pulse quality; rather, it is the area of the action diagram. All of the areas are seen to be about the same, approximately 2 eV-µs. Some obvious distortion of the extracted pulses takes place by non-linear transformations (the action diagrams are not parallelograms), but it appears that a large fraction of the collection could still be time-focused to within a longitudinal emittance of 2 eV-µs. By adjusting the focal length of such a lens, an energy spread of 1

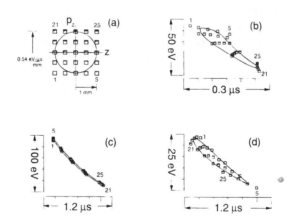

FIGURE 17
Action diagrams from computations of energies and time of ions extracted from a Paul trap. The particles are labeled from 1 to 25, as shown on the action diagram for the macromotion in the trap (a). (b) Action diagram that results at a distance of about 700 mm from the trap center when the RF phase at the initiation of extraction was 197°. (c) For a phase of 94°. (d) For a phase of 215°. The dashed outlines in (a), (b), and (c) are for the circle in the action diagram for the macromotion. Details of the trap operating parameters and the extraction process are given in text.

eV could be obtained for a pulse duration of 2 μs. Alternately, a time focus to 0.1 μs could be obtained at the expense of an energy spread of about 20 eV.

This example shows clearly the need to know the actual action diagrams for a particle collection in a Paul trap if one is to tailor an extraction system with time-refocusing for a specific use. Of particular concern is the longitudinal action diagram which determines the focal lengths of the time lenses that are needed.

VI. EXAMINATION OF TRAPPED IONS BY PULSED EXTRACTION

A direct examination of the longitudinal action diagram of a beam pulse extracted from a Paul trap would require simultaneous measurement of the time of arrival of particles at a detector at a given position, together with either the momentum and position of particles in a beam pulse at a particular time or of the energy. In principle, a measurement of the Doppler shift of radiated light as a function of axial position of the ion would give the necessary information. However, this would depend on the availability of easily excited visible radiation from the ions, and is seldom feasible for the type of ions and the velocities and particle densities that are typical for a beam pulse from a Paul trap.

On the other hand, simultaneous measurement of energy and time of arrival is also typically very difficult, although in principle it can be done by passing the beam pulse through an energy filter and observing the time of arrival of the filtered ions (Figure 18). Such an observation would give the density along a horizontal slice of the action diagram at a particular energy. By varying the energy that is filtered, a full view of

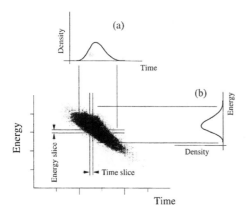

FIGURE 18
The concept of examination of an action diagram by the analysis of slices. (a) Representation of analysis by horizontal slices obtained from an energy filter. (b) Representation of analysis by vertical slices obtained by a time gate.

the particle density of the action diagram would be obtained. However, energy filtering is not easy. Filtering by momentum selection, using a magnetic or electric bending of the particle trajectories causes the transverse actions of the particles to be folded into the selection process. Energy gating by retarding electric fields would introduce severe transverse defocusing of the beam pulse.

The alternative of time gating of the beam pulse and observation of the energy spread of the resulting vertical slice of the action diagram is not accomplished readily; it is very difficult to turn on and off the gating electric fields quickly enough in order to not introduce energy changes that vary with the time of transit of the particles.

The only observation that is relatively easy to make, to the accuracy required for detailed examination of the action diagram of an ion pulse from a Paul trap, is that of TOF. By accelerating the ions to sufficient energy, the ions can be detected with efficiencies approaching 100% and their times of arrival can be measured to within nanoseconds. Furthermore, the speed of modern digital oscilloscopes allows the shape of a single ion pulse to be recorded accurately.

The observation of the shape of a beam pulse is equivalent to observing the density of the action diagram from a vertical viewpoint (Figure 19). Information for an orthogonal point of view is of course lost. If, however, by transporting the beam pulse to a new location, the action diagram could be transformed so that the time spread was due almost entirely to the initial energy spread, then the observation of the pulse shape becomes an observation of the original action diagram from a horizontal point of view (equivalent to Figure 19c).

The problem is analogous to that of using an X-ray beam to probe a structure using axial tomography scanning. In this process, the density of a structure is probed by passing a beam through it in various directions. By using an inverse spatial Fourier transform, the transmission of the beam in various directions can be used to compute a three-dimensional density map of the structure. The process is called slice-projection in radar imaging or, in its more familiar form, computerized axial tomographic (CAT) scanning in medical imaging.

To apply this technique to the examination of the action diagram of a beam pulse, a transport system is needed that transforms the action diagram into a new one in which the time coordinate can be made to depend in a variable way on the energy-time (or momentum-displacement) coordinates of the original diagram. The effect of such a transformation would be equivalent to looking at the density of the action diagram from arbitrary points of view (Figure 19, b and d). In principle, the views from all possible directions give all the information needed to discern the full three-dimensional shape of the density of the action diagram at all energy-time points.

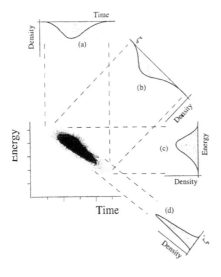

FIGURE 19

Analysis of an action diagram by projection into planes from various points of view in the action diagram. (a) The result of plotting the integrated density through the diagram as a function of time (i.e., looking in the direction of the energy axis). (c) The result of plotting the integrated density as a function of energy. (b) and (d) The results of plotting the density as a function of linear combinations of the energy and time coordinates, indicated as ζ and ζ'.

Such a situation has been seen to occur for the extraction of an ion pulse from a Paul trap simply by applying an extraction pulse at different phases of the RF (Figure 17). At one extreme of the RF phase, the time distribution of the action diagram is due primarily to the momentum distribution of the action diagram for the macromotion at the trap center; at the other extreme, it is due primarily to the spatial distribution. Therefore, information required for discerning the action diagram at the center of the trap is available simply by observing the shapes of the extracted beam pulses as the RF phase at extraction is varied.

The problem that remains for the implementation of this technique is to decode the information in these pulse shapes. The technique used by Lunney[7] was to assume a Gibbs distribution at a given temperature for the particle distribution in the macromotion action diagram at the trap center. The action diagram was then divided into pixels, typically a 21 × 21 array, with each pixel weighted according to the Gibbs distribution. The action diagram was then given the RF distortion corresponding to the particular RF phase at extraction. Computer simulation of the transfer of the pixels to the detector was then carried out. By using the weight of each pixel, the time of arrival and the shape of the beam pulse at the detector could be simulated. The temperature of the Gibbs distribution was then varied until a satisfactory match to the experimental results was achieved.

The extraction and beam transport system was designed to give maximum sensitivity of the beam pulse shape to the momentum and displacements expected in the action diagram at the trap center. A schematic diagram of the apparatus, with typical operating potentials, is shown in Figure 20.

A comparison of the calculations with typical pulse shapes observed is shown in Figure 21. The only free parameters used in fitting the simulations to the observations, in order to normalize the height of the peak of the simulated pulse, were the temperature of the Gibbs distribution and the total weight of the action diagram. The accuracy of the computer simulation of the particle extraction and transport is shown by the degree to which the time of the peak of the beam pulse agrees with the experimental results (estimated to be within 0.1 μs in a flight time of 12 to 14 μs). The validity of modeling the action diagram at the trap center using Gibbs distribution is demonstrated by the accuracy with which the shape of the rising edge of the beam pulse is simulated. (The trailing edge of the beam pulse suffers from lower energy ions resulting from scattering off the grids used in the transport system. The different contributions from this scattering at different RF phases are due to variation of the transverse focusing powers of the extraction electrode.)

The temperatures of various collections of potassium and sodium ions were derived in the fashion described above. It was found that with

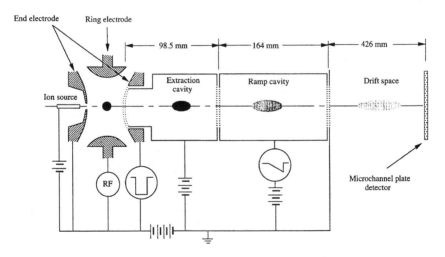

FIGURE 20
Schematic of the trap loading, operation, and extraction system for diagnosing the longitudinal action area of ion collections in a Paul trap. Typically, the trap was biased to 800 V, and the extraction potential applied to the RHS end-cap electrode was 200 V. The extraction end-cap electrode was formed from a stainless steel mesh to preserve the quadrupole field within the trap. The electric potentials within the cavities were maintained by covering the openings with planar grids. The separation between adjacent cavities was 1 mm.

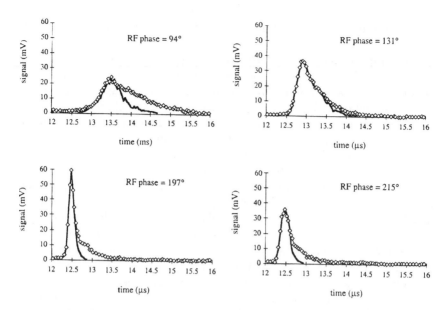

FIGURE 21

Observed pulse shapes (□) and computer simulations (solid lines) for ion collections extracted from a Paul trap at different RF phases for initiation of extraction. The measurements were obtained by summing the results of 1000 pulses analyzed by a digital oscilloscope for each of the RF phases shown.

fixed trap operating parameters, the temperature of the macromotion of small ion collections in a Paul trap, under buffer gas cooling, varied with approximately the two thirds power of the number of ions delivered to the detector. The results are summarized in Figure 22.

The results shown in Figure 22 show that the temperature of small collections of ions in a Paul trap can indeed be very low. The overall efficiency with which ions in the trap were detected is estimated, very crudely, to be about 5%. Figure 22 indicates, therefore, that with 10^4 ions in the trap, the temperatures were about 0.25 eV for potassium and 0.15 eV for sodium. At these temperatures, about 90% of the ions are contained within an action area of about 2 eV-μs.

The decrease in temperature with the number of ions in the trap shown in Figure 22 is particularly encouraging. For 10^3 ions in the trap, it is seen that the temperature approaches that of the buffer gas itself. Because the action area is proportional to the temperature of the collection, the area will be on the order of 0.2 eV-μs. Such an action area would rival those obtained from pulsed ionization by lasers, one of the preferred methods for producing ions for TOF MS.

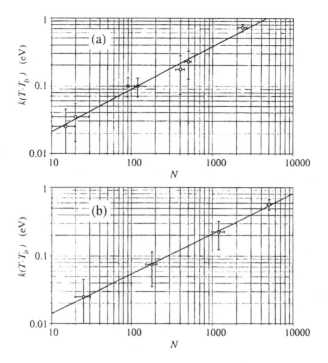

FIGURE 22
Derived temperatures of the Gibbs distributions of the macromotions in a Paul trap for different numbers of ions (N) observed at the detector. (a) Potassium ions. (b) Sodium ions. The ordinate is the difference between the derived Gibbs temperature, T, and the temperature of the buffer gas, T_b (in this case, room temperature or 0.025 eV).

VII. CONCLUSIONS

The possibility of efficient collection of beams of charged particles in an electromagnetic trap and their subsequent cooling to very small phase space volumes is very important in experimental science. Such a possibility permits the separation of charged particle production from charged particle confinement. Furthermore, the possibility of extracting the collection as a very accurately defined volume in phase space greatly enhances the applicability of other techniques, such as TOF MS, for example, which depend on the availability of such collections. Electromagnetic traps no longer should be considered as mere benchtop toys, but as essential components of sophisticated apparati that can be engineered to accomplish a multitude of tasks in physics and chemistry.

REFERENCES

1. Becker, St.; Bollen, G.; Kern, F.; Kluge, H.-J.; Moore, R.B.; Savard, G.; Schweikhard, L.; Stolzenberg, H. *Int. J. Mass Spectrom. Ion Processes.* 1990, 99, 53.
2. Lunney, M.D.N., Ph.D. thesis, McGill University, Montreal, Canada, 1992.
3. Michael, S.M.; Chien, M.; Lubman, D.M. *Proc. 41st ASMS Conf. Mass Spectrom. Allied Topics.* San Francisco, CA. 1993, 942.
4. Todd, J.F.J.; Freer, D.A.; Waldren, R.M. *Int. J. Mass Spectrom. Ion Processes.* 1980, 36, 185.
5. Todd, J.F.J.; Freer, D.A.; Waldren, R.M. *Int. J. Mass Spectrom. Ion Processes.* 1980, 36, 371.
6. Crawford, J.E.; Buchinger, F.; Davey, L.; Ji, Y.; Lee, J.K.P.; Pinard, J.; Vialle, J.L; Zhao, W.Z. *Hyperfine Interactions.* 1993, 81, 143.
7. Moore, R.B.; Gullick, S. *Phys. Scripta.* 1988, T22, 28.
8. Rouleau, G., Ph.D. thesis, McGill University, Montreal, Canada, 1992.
9. Lunney, M.D.; Moore, R.B. *IEEE Trans. Mag.* 1991, 27, 4174.
10. Knight, R.D.; Prior, M.H. *J. Appl. Phys.* 1979, 50, 3044.
11. Dehmelt, H.G. *Adv. Atom. Mol. Phys.* 1967, 3, 53.
12. Dehmelt, H.G. *Adv. Atom. Mol. Phys.* 1969, 5, 109.
13. Baril, M.; Septier, A. *Rev. Phys. Appl.* 1974, 9, 525.
14. Lunney, M.D.N.; Buchinger, F.; Moore, R.B. *J. Mod. Opt.* 1992, 39, 349.
15. Vedel, F.; Vedel, M.; March, R.E. *Int. J. Mass Spectrom. Ion Processes.* 1991, 108, R11.
16. Todd, J.F.J.; Waldren, R.M. *Int. J. Mass Spectrom. Ion Phys.* 1979, 29, 301.
17. Waldren, R.M.; Todd, J.F.J. *Int. J. Mass Spectrom. Ion Phys.* 1979, 29, 315.
18. Waldren, R.M.; Todd, J.F.J. *Int. J. Mass Spectrom. Ion Phys.* 1979, 29, 337.
19. Mather, R.E.; Todd, J.F.J. *Int. J. Mass Spectrom. Ion Phys.* 1979, 31, 1.
20. Waldren, R.M.; Todd, J.F.J. *Int. J. Mass Spectrom. Ion Phys.* 1979, 31, 15.

Author Index

A

Amy, J.W., 45

B

Beauchamp, J.L., 165
Bier, M.E., 453
Bollen, G., 258–260
Bortolini, O., 145
Bowers, W.D., 201
Brauman, J.I., 165
Brodbelt, J.S., 205, 206
Brown, L.S., 246

C

Cooks, R.G., 3, 16, 20, 36, 38–43, 64–
    65, 67–68
Cornell, E.A., 247, 255
Cotter, R.J., 218
Cox, K.A., 3, 8, 20, 25, 26, 33, 36
Creaser, C.S., 171

D

Davey, L., 280
Debye, P., 154
Dehmelt, H.G., 287
Dunbar, R.C., 165, 171

E

Edmonds, C.G., 140

G

Gabrielse, G., 246–248

Gerz, Ch., 246, 256
Glish, G.L., 89
Goeringer, D.E., 218, 228
Gullick, S., 282, 283

H

Hemberger, P.H., 45, 180, 188, 200
Hillenkamp, F., 217, 226
Hughes, R.J., 53, 166, 168

J

Julian, R.K., Jr., 45

K

Kaiser, R.E., Jr., 9, 13, 14, 17–19, 45
Karas, M., 217, 226
Kluge, M.-J., 258–260
Kofel, P., 51

L

Lammert, S.A., 45, 187, 200
Londry, F.A., 30
Louris, J.N., 45, 169
Lunney, M.D.N., 263, 284, 291, 298

M

March, R.E., 53, 66, 168
Mather, R.E., 294
McLuckey, S.A., 89
Mikami, N., 173, 175, 179
Moore, R.B., 263, 282–284
Morand, K.L., 45

R

Rouleau, G., 282, 284

S

Schlunegger, U.P., 67–69, 85
Schwartz, J.C., 3, 89, 140
Smalley, R.E., 56
Stafford, G.C., Jr., 10, 45, 221
Stephenson, J.L., Jr., 163
Syka, J.E.P., 10, 45

T

Tan, J., 247, 248
Todd, J.F.J., 10, 265, 283, 294
Traldi, P., 145

U

Uechi, G.T., 171

V

Van Berkel, G.J., 89
Van Dyck, R.S., Jr., 247, 253, 255
Vargas, R.R., 205, 217
Vedel, F., 237, 292

W

Waldren, R.M., 294
Watson, C.H., 192
Werth, G., 237, 256
White, J.U., 192
Williams, J.D., 3, 38–41, 182, 200, 201
Wilsdorf, D. 256

Y

Yost, R.A., 163, 205, 217

CHEMICAL INDEX

A

Ar$^+$, 83
Au$^+$, 65
Acetaldehyde, 194
Acetic acid, 96–98
Acetophenone, 182, 187
Adenine, 137–139
Air, 93, 226
Alkyl benzenes, 155–157, 171
Allyl bromide, 192
Ammonium salts, 95
Angiotensin I, 107, 109
Anthracene, 69, 215–216
Antiproton, 257
Arginine, 129, 131–132
Argon, 55, 65, 196–198, 220

B

Br$^+$, 212, 214
Benzaldehyde, 171
Benzene, substituted, 157–158
Benzoyl cation, 180, 182–183, 186–187, 189–190
Biomolecules, 105–106, 111, 124, 126, 128, 200, 206–207, 228, 265
Biopolymers, 95, 99, 105, 118–119, 134, 139
Biphenyl, 155
Bovine insulin, 122, 129
Bradykinin, 20
iso-Butylbenzene, 171–172
n-Butylbenzene, 165, 170–172
sec-Butylbenzene, 171–172
tert-Butylbenzene, 171–172

C

C_3^+, 222–223
C_4^+, 222–223, 254
C_5^+, 222
$CH_3^{\cdot}$, 172
CH_4, 81, 83–84
CH_3OH, 7
CO, 238
CO_2, 165–166, 192–193, 196, 207
C_2H_5, 172
$C_3F_5^+$, 26
C_6H_5Cl, 173–175
$C_6H_5CO^{+\cdot}$, 180
C_6H_5OH, 176
Cr$^+$, 212–213
Cs, 259
^{133}Cs, 259
Cs$^+$, 6–9, 11–14, 16, 24, 37, 106, 281
Cs(CsI)$_{31}$, 55
CsF, 26, 37–41, 44
$Cs_5F_8^+$, 43
$Cs_7F_6^+$, 39
$Cs_8F_7^+$, 39
$Cs_9F_8^+$, 28, 44
CsI, 5, 7, 12, 16–18, 24, 30, 35–37, 42–43, 65
Cs_2I^+, 42
$Cs_3I_2^+$, 42
$Cs_4I_3^+$, 42
$Cs_{10}I_9^+$, 11
$Cs_{38}I_{37}^+$, 12
(CsI)$_{122}$Cs$^+$, 17
(CsI)$_n$Cs$^+$, 11, 13, 17
^{63}Cu$^+$, 220-1
^{65}Cu$^+$, 220-1

Carboxyl ion, 229
Chlorobenzene, 173
m-Chloronitrobenzene, 151–152
o-Chloronitrobenzene, 151–152
p-Chloronitrobenzene, 151–152
Chromium, 220
Collagen IV, 230
Copper, 220, 243
Coproporphyrin-I tetramethylester,
    96, 97
12-Crown-4, 210–212
15-Crown-5, 212, 214
1,4-Cyclohexadiene, 187
Cytochrome c, 106, 121–122

D

D·, 119
Diacetone alcohol, 168
1,6-Diaminohexane, 129, 133
Digitoxigenin, 212, 214
Diglyme, 194–200
2,5-Dihydroxybenzoic acid, 226, 229
Dimethylamine, 120–122, 131–133
p-Dioxane, 176–177
Dithiothreitol, 24, 26

E

Egg albumin, 18–19
Electron, 257
Entactin, 230
2-Ethanethiol, 168
Ethanol, 168
Etioporphyrin-III, 100

F

Fe⁺, 212–213, 216
⁵⁶Fe⁺, 224–225
FC-43, see Perfluorotributylamine

G

Gallium hexafluoroacetylacetonate,
    192
Glu-fibrinopeptide B, 111, 113–114
Glycerol, 7–8, 37

Gold, 206, 243
Gramicidin S, 28, 30, 37–44, 212, 215
Graphite, 222–3

H

H·, 119
³H, 238, 257
H₂O, 7, 32–33, 37, 69, 84, 120, 138,
    139, 177
He⁺, 242
³He, 238, 257
Helium, 55, 64–65, 67–69, 73–75,
    78–79, 83–84, 92, 99, 129, 169,
    180, 184–185, 192, 196–199,
    209, 219–221, 250, 252, 281
Hematin, 103–104
Heparan sulfate proteoglycans,
    230
Heptacarboxylporphyrin-I
    heptamethylester, 97
n-Heptane, 67
Hexacarboxylporphyrin-I
    hexamethylester, 97
1,6-Hexanediamine, 122–123
Histidine, 129
Horse angiotensin, 25, 27
Horse skeletal muscle myoglobin,
    106–107, 120, 134
Human growth hormone, 117–118
Human hormone releasing factor,
    116
Hydrocarbons, 69
Hydrocarbons, polycyclic, aromatic,
    215
2-Hydroxyethanethiol, 168

I

Iron, 215, 220
Isopropanol, 168
Isopropylbenzene, 155

K

K⁺, 210–212
³⁹K⁺, 294
KBr, 210, 214–215

L

Laminin, 230
Leucine-enkephalin, 68, 110, 207, 212, 215, 228–230
Lysine, 129, 131–132

M

Matrigel, 230–231
Melittin, 131–133
Mesoporphyrin-IX dimethylester, 96–97
Met-(O)-enkephalin, 110
Metal salts, 95
Methanol, 95–98, 120, 176, 194, 226–227
Methyl acetate, 96
Methyl propionate, 96
Methylene chloride, 95

N

N_2, 83, 93, 196–198, 219–220, 238
N_2^+, 74, 77–79
Na, 155
Na^+, 135–136
NaCl, 166, 215
$N(CH_3)_3$, 176
Ne, 65
NH_3, 32–33, 37, 177–178
$^{15}NH_3$, 238
$^{14}NDH_2$, 238
Naphthalene, 212–213
Neuromedin U-8, 115
Nickel, 220, 253
Nicotinic acid, 226–227
Nidogen, 230
Nitrobenzene, 147–148
Nitrogen, 228, 233
Nucleotides, 139

O

O_2^+, 69, 83–84
Octaethylporphyrin, 97–98
Oligonucleotides, 91, 119, 134–137, 139

P

PO_3^-, 137
Pentacarboxylporphyrin-I pentamethylester, 97
Pentacene, 215–216
Peptides, 91, 95, 105, 110–112, 116, 124–125, 127, 129–130, 132, 134–135, 137
Perfluorokerosene, 166
Perfluoropropene, 192
Perfluoropropylene, 171
Perfluorotributylamine, 5, 24, 27, 64, 67, 106, 146
Phenol, 175, 179
Phenol-H_2O, 179–180,
Phenol-NH_3, 177, 179
Phenol-p-dioxane complexes, 179–80
Phenol-trimethylamine, 176–177
Phenoxy radical, 176–177
Phenyl, 212
Phosphonium salts, 95
Polydeoxy-adenoic acid, 135–136
Porphyrin esters, 97
Porphyrins, 91, 95–97, 99, 101, 110
Positron, 257
Potassium, 227, 299–301
Propane, dimer, 168
1-Propanethiol, 168
2-Propanethiol, 168
2-Propanol, 166–8
Proteins, 91, 94, 105, 110–111, 116, 129–130, 132, 134–136, 139
Proton, 124, 130, 135, 257
Pyrene, 69
Pyridines, alkyl-substituted, 156–157

R

^{84}Rb, 260
RbI, 12, 14
$Rb_5I_4^+$, 12
$Rb_9I_8^+$, 12
Radionuclides, 264–265, 280
Renin substrate, 31, 33–34, 108, 119
Rhenium, 258
Rhodamine, 176

S

Sinapinic acid, 226, 231
Sodium, 227, 229, 231, 299–301
Somatostatin, 26
Spiperone, 225–232
Steroids, 95
Styrene, 155
Substance P, 24–26, 28–29, 32–34, 45
Sucrose, 207
Sugars, 137–139
Sulphonated azo dyes, 95

T

Tantalum, 206
Tetraphenylporphine, 212, 214
Thioglycerol, 7–8
Toluene, 69, 95–98
Transition metal complexes, 95
Triethylene glycol dimethylether, 210–212
Trifluoroacetic acid, 95–97, 226–227
Trifluoromethylbenzene, 150–151
o-Trifluoromethylbenzonitrile, 149

p-Trifluoromethylbenzonitrile, 147, 149–150
Triglyme, 211
Trimethylphenylammonium chloride, 228
Tritium, 257

U

Uroporphyrin-I octamethylester, 96–97

V

Vanadyl etioporphyrin-III, 100–102

X

Xe, 27, 65, 281
Xe⁺, 26, 69, 147–148, 281–283
XeCl, 176–177, 180, 182, 187
o-Xylene, 155
p-Xylene, 155

Z

ZnSe, 192

SUBJECT INDEX

A

a_z, definition, 10
Absorption spectrum, 179
Accurate mass assignment, 6
Action diagram, 266–270, 272–277,
  279–280, 286–289, 294–296
  analysis, 298
  area, 295
  axial, 280
  computer simulation, 293
  elliptical, 276–277, 287
  energy time, 274
  longitudinal, 275, 296
  modeling, 299
  parallelogram, 276
  rectangular, 273
  RF distortion, 289–290
  total weight, 299
Adducts
  formation, 211, 228
  ion, 133
  sodium, 231
Advantages of MS^n approach,
  101
Amplitude reaction, 249
Analog-to-digital converter (ADC),
  22, 82
Analyte-to-matrix ratio, 226
Apex isolation, 194, 211
Artifact peak, 12
ASGDI/ion trap system, 92
ASGDI source, 92
Asymptotic injection, 56, 66
Atmospheric pressure ionization
  (API) sources, 265

Atomic polarization, 153
Atomic standard of mass, 260
Attachment reaction, 212
Avogadro's constant, 261
Auxiliary AC voltage, 15, 151,
  182–183
  frequency, 185
  phase, 184
Auxiliary RF amplitude, 185
Auxiliary RF potential, 6, 7, 94, 99,
  212
Axial
  distribution, 181
  excursion, 199, 200
  injection, 56, 68
  ion motion, 291
  kinetic energy, 185
  modulation, 15, 147, 151, 225
  oscillation, 244, 249, 252
  oscillation damping, 282
  position, 185
  resonance, 255
  tomography scanning, 297
Azimuthal quadrupole field, 282

B

B, coefficient,
B_{eject}, 15, 17
$-_r$, 9
$-_z$, 9, 16, 17, 225
Background pressure, 129
Bath gas, 99, 113, 128
Bath gas dampening, 106
Bath gas pressure, 169
Beam defocusing, 272

Beam emittance, 268
Beam focusing, 272
Beam optics, 267, 268
Beam profile, 72
Beam pulse quality, 275
Beam pulse shape, 297–298
BE/ion trap, 67
BE/quistor/quadrupole, 67
Beta-decay, 257
Biochemical research, 111
Biological molecules, desorption, 212
Biomolecular reaction, 126
Biomolecules, 228, 265
    mass resolution, 4
Black holes, 225
Boxcar integrator, 82
Broad-band kinetic excitation, 115
Bromine-bound dimers, 214
Buffer gas, 99, 113, 128, 181, 197, 219, 221, 251, 281, 293, 300
    collision,
    pressure, 129, 168, 220–221, 225
Bunched ions, 283,
Bureau International des Poids et Mesures at Sèvres, France, 260

C

Calibrant ions, 37, 40
Calibration, 39, 41, 166
Calibration curve, 40
Calibration standard
    external, 37
    FC-43, 146, 166
    internal, 37–38
Canonical coordinates, 286
Capillary electrophoresis, 110–111
Central beam velocity, 271
CERN, 258, 264, 280, 282–283
Channel-plate detector, 254
Charge
    exchange reaction, 215
    separation, 126
    state distribution, 121, 124
    state information, 116
    transfer chemistry, 90

Charged particle beam optics, 267–268
Chemical noise, 99
CI, 233
CIT/photoionization/
    photodissociation apparatus, 173
Cluster complex, 123
Cluster ion
    formation, 128–129
    internal energy, 128
    kinetic energy, 128
    lifetime, 129
Cluster ions, 12, 16–17, 35, 37, 41, 123, 128, 175
Cluster reaction, 123–124
Clustering, 121–123, 128–129, 131, 133–134
Clustering degree, 128
Co-injection method, 6
Collection device, 263
Collection of free particles, 266
Collection of particles, 270
Collision cell, 66–67
Collision activation
    interface, 99, 101
    ion trap, 103, 111
    random noise, 103
Collision-induced dissociation, 66–67, 99, 113, 164, 228
    high energy, 171
    low energy, 171
    spectrum, 170
Collisional
    activation, 100, 103, 111, 113, 115, 117, 124, 133, 136, 139
    cooling, 168, 209
    deactivation, 196
    stabilization, 127
    thermalization, 67
Compensation electrodes, 245, 248
Conversion dynode, 94
Correction electrodes, 244
Collection efficiency, 282–284
Complex formation, 211
Computerized axial tomographic (CAT) scanning, 297
Consecutive ion isolation, 194
Continuous beam injection, 280
Coulomb repulsion, 281

Coulomb repulsion, internal, 136
Cylindrical ion trap (CIT), 165,
    173–174, 248
Cylindrical ion trap,
    photodissociation, 173
Cyclotron frequency, 239, 241, 249,
    255, 257
Cyclotron resonance, 260

D

Damping collisions, 209
Damping gas, 106
Data acquisition scan speed, 25
Daughter ions
    axial modulation, 151
    mass spectra, 68
DC pulse, rapid, 200
De Broglie frequency, 260
Deceleration lens, 61
    system, 60, 284–285
Defocusing elements, 273
Desorption, 214–215
Detection
    destructive, 254
    limits (ES/MS), 97
    nondestructive, 252
    of resonances, 252
Dielectric constant, 126, 154
Digitizing rate, 107
Dipolar polarizability, 157
Dipole moment, 153, 158
Displacement, 271–272
Dissipative processes, 55
Dissociation
    rate constant, 166–167
    rates, 165
    threshold, 165
Doubly charged peptides, 112
Dye laser, 173

E

Effective ion temperature, 84
Einzel lens, 63, 206
Electrochemistry, 90
Electrode imperfections, 106,
    243

Electrodes
    cylindrical, 173–174
hyperboloidal, 282
    reduced size, 7, 11
Electron
    multiplier, 94
    photodetachment spectrum, 166
Electronic polarization, 153–158
Electrospray (ES), 89
    apparatus, 92, 139
    detection limits, 97
    interface, 92, 101–102
    interface fragmentation, 99,
        101–103, 113
    ionization, (ESI), 5–7, 89–90, 105
    mass spectrum, 25, 27, 95–98, 102,
        107–110, 114, 120, 122, 135–136
    solvent, 96
    solvent stream, 98
    source, 93
Electrospray
    desolvation/ion trapping, 124
    ion trap mas spectrometry
        (ES/ITMS), 95, 99, 134, 139
    ion trap (ES/IT), 90–94, 110–111
    mass spectrometry (ES/MS), 95,
        105
Electrosprayed peptides, 110
Electrosprayed proteins, 110
Electrostatic deceleration, 55, 67
Emittance, 268, 271, 276
    diagram, 268, 276
End-cap electrode, 11, 94, 187, 208,
    245, 254, 278
    flat, 248
    hyperboloid profile, 11
    separation, 283
    Splitting, 282
Energy
    deposition, 99
    diagram, 126, 128
    diagram, hypothetical, 128
    surface, 126
Enhanced mass resolution, theory,
    30
Equations of motion, 266
Equipotential contour lines, 71
Excitation period, 99

External
 calibration, 37
 ion source, 6, 64, 233
Extraction efficiency, 81
Extraction of particles, 291

F

Fast DC pulse, 187
Fast DC pulse probe, 186–191,
 200
Femtomole levels, 111
Fiberoptic laser probe interface, 169,
 207–208
Fiberoptic probe, 170–171, 207–208
Finnigan, 5
 electrospray/ion trap, 92
 software, 146
Finnigan MAT
 Corp, 94
 ES/IT, 92
 ITMS, 91, 207, 218, 233
Focusing elements, 273
Fourier transform, 187–188, 297
Free energy, 127
Frequency
 domain data, 189–190
 reduction, 11
 shift, 243, 249
FTICR, 4, 5, 165, 207, 218
Full width at half-maximum
 (FWHM), 5, 22–24, 40,
 181–182, 186
Fundamental
 axial secular frequency, 16, 151,
 187–188
 secular frequency, 187

G

Gas chromatograph, Varian 3400,
 171
Gas phase basicities, 127
Gaussian distribution
 spatial distribution density, 287
GC/MS, 5
GC/MS/MS, 171
Gibbs distribution, 285–286, 298

Gibbs distribution temperature, 299,
 301
Glow discharge source, 92

H

Halide ions, desorption, 212
Hard sphere collision cross-section,
 129
Harmonic
 oscillation, 245
 oscillator, 241
Hartree-Fock multi-electron theory,
 259
H/D exchange, 119
Helium pressure, 209
Hexapole field contribution,
 188
High mass
 compounds, 4
 measurement accuracy, 4, 6
 resolution, 3–6, 21, 116
 resolution, application, 32
 resolution zoom, 107, 117, 135
High order multipole components,
 284, 293
High performance liquid
 chromatography, (HPLC),
 99
High performance mass
 spectrometer, 91
High precision mass spectrometry,
 237
High precision trap, 258
High resolution
 ion isolation, 31–32, 34, 118–119
 mass analysis, 24
High resolution mass spectrometry
 experimental, 29
 simulated, 29
High resolution mass spectrum, 26,
 28
High resolution MS/MS, 5
High resolution parent
High resolution parent ion selection
Higher-order field, 187
Highly-charged proteins, 136
HPLC/ES/MS/MS, 103–104

Hybrid
    mass spectrometer, 66
    MS system, 218
Hyperbolic ion trap electrode
        structure, 282
Hyperboloid of revolution, 240

I

In situ ion formation, 232
Increased resolution, 14
Infrared (IR) laser, 166
Injected ions, See also Ion injection
    mass selected, 51, 53, 66, 85
Injection
    efficency, 59, 73
    fragmentation, 99, 101–103
    of mass-selected ions, 51, 53
    optics, 72, 80
    RF voltage amplitude, 102
    situation, 57
    transmission, 75
Integrated ion signal, 108
Interface
    CID spectrum, 113
    fragmentation, 101–103, 113,
    lenses, 101–102
Internal calibrants, 6
Internal calibration, 37
Ion
    activation, 111
    atom collision, 55
    beam profiles, 71
Ion cloud, 181, 185–187, 265, 291
    bunch, 283
    dynamics, 181–191
Ion cooling, 221
Ion current profiles, 98, 104
Ion cyclotron resonance cells, 249
Ion desorption/ionization, 207
Ion distribution, 184–185
Ion ejection, 232
Ion energy
    axial, 185
    deposition, 170
Ion fragmentation, 30
Ion frequency determination,
        186

Ion injection, 5, 51, 53, 62, 69, 93–94,
        102, 113, 276, 279, 283
    continuous beam, 280–281
    deceleration system, 284
    effects of RF phase, 283
    energy, 64–65, 75–77
    lens system, 232
    low energy, 85
    on axis, 233
    optics, 72
Ion/induced dipole, 126
Ion-ion interaction, 115
Ion/ion reaction, 125–126
Ion isolation, 94, 187
Ion kinetic energy (ies), 182
    axial, 185
    diminution, 209
Ion/molecule
    collision, 55
    complex, 123
    interactions, 150
    reaction, 83, 92, 119, 124–126,
        130–132, 134, 207, 213
    reactivity, 131
Ion motion, 183, 290
    damped, 77–78
    equation, 239–240
Ion optics, 70, 93
Ion polarizability, 145, 153–156,
        158
Ion reaction, 209
Ion spacing, precise, 44
Ion stopping distance, 79
Ion storage capacity, 105
Ion structure, 164
Ion tomography, 180, 297
Ion trajectory
calculations, 71
    Penning trap, 241, 252
    simulation, 284
Ion trap
    capabilities, 105
    commercially available,
        106
    electronics, 182
    Finnigan MAT, 7, 15
    half-size, 13
    imperfections, 243

mass spectrometer, *see* Mass
  Spectrometer, ITMS
mass spectrometry, 180, 192, 206,
  217
operating parameters, 8
parameters, 243
Penning, 62–63, 237–238, 240, 244,
  247, 249, 251, 253–255,
  257–258, 260, 292
pressure, 80
pulse pressurized, 66
Purdue University, 7
quarter-size, 13
reduced size, 68
scan program, 19
selectivity, 226
sensitivity, 226, 232
Ion trapping efficiency, 220,
  223
Ionization potential, 174
Ions, multiply-charged, 5, 25, 30,
  91–92, 116, 118, 126–127
Ionspray, 5
IR laser desorption, 207
IR laser desorption, pulsed, 207
Iso-_ lines, 16
Iso-_z lines, 15
Isobaric ions, 27, 147
ISOLDE, 258, 264, 280, 282–283,
  285
ISOLDE mass separator, 258
Isomer differentiation, 165, 168,
  171–172, 213
Isomeric ions, 147, 151–153
Isotopes composition, 117,
  220–221
Isotopic analogues, 168
ITMS™ software, 37, 103, 146
ITSIM, 27, 191

K

Kilogram standard, 260
Kinetic control, 127
Kinetic energy
  distribution, 72
  reduction, 55, 209
Kinetic scheme, 123

L

Lagrange-Helmholtz equation, 58, 70
Lambshift, 256
Lambshift measurement, 238
Laplace equation, 240, 245
Laser
  beam, 208
  continuous, 165
  desorbed reagent ion, 216, 220–221,
    224, 228, 280
  desorption, 5, 6, 205–207, 209–210,
    213, 216–218, 226, 233
  desorption event, 210, 220
  desorption/ion injection/ITMS,
    232
  desorption ionization (LDI), 217
  desorption, pulse, 169, 210
  frequency, 176
  frequency scanning, 176
  induced dissociation, 168, 180
  irradiance time, 194, 196
  microscopy, 233
  photodissociation, *see*
    Photodissociation
  photolysis, 174
  power, 171
  probe, 186, 200
  pulse energy, 219, 228
  pulse synchronization, 222
  pulsed, 165, 167, 169
  spectroscopy, 264
  trigger, 222
Lasers
CO$_2$ continuous wave, 165–166,
    192–193, 207
dye, 171, 173–174, 176
  infrared, 206
neodymium-YAG, 169, 171, 174, 176
  nitrogen, 218–219, 228, 233
  XeCl excimer, 176–177, 180, 182, 187
LDI, 220
LDI/ITMS, 217, 219
LDI mass spectrum, 221, 226
LDI/MS/MS, 217–218
LDI system, external, 232
Lens system, 92
Linear accelerator, 281

Liouville theorem, 265
Lissajous' motion, 188
Longitudinal emittance, 269, 275, 278
Lorentz force, 240
Los Alamos National Laboratory, 7
Low-mass cutoff, 101, 113, 224–225

M

Macromotion, 288–289
Macromotion trajectory, 289
Magnetic field fluctuations, 247
Magnetic field gradient, 247
Magnetron frequency, 241
Magnetron oscillations, 244, 249
MALDI, 228, 232
    mass spectrum, 19, 227–231
    MS/MS daughter spectrum, 229
    optimum setting, 228
    sample preparation, 226
Mass
    accuracy, 5, 38, 106
    calibration procedure, 146, 166
    displacement, 147, 149–150,
        152–153, 155–158
    doublets, 256
    measurement accuracy, 4, 35, 38,
        43
    range extension, 5–8, 10–12, 14–18,
        21, 107, 147, 209, 212, 225
    resolution, 22, 107, 238
    selected ion injection, 51, 53, 66, 85
    selected ions, 51, 53
    selection, 58
Mass-selective
    axial instability, 148
    instability scan, 7
    instability scanning method,
        220
Mass shift, 6, 28, 35–36, 43, 147
    effect,
    negative, 147
    positive, 147
Mass spectrometer
    high performance, 91
    ITMS™, 4, 90, 146–147, 169, 171,
        192, 207
Mass spectrum of CSI clusters, 17–18

Mass spectrum of RbI, 14
Mass window, 113
Massachusetts Institute of
    Technology, 243
Mathematic formalism, 265
Mathieu equation, 78
    coefficient, 10
    parameters, 10
Matrix-assisted laser desorption
    ionization (MALDI), 217, 226
Mattauch Herzog mass
    spectrometer, 259
Maximum axial excursion, 35,
    185–186
Maximum excursion, 35, 186
Membrane probe, 69
Metal ion, reactions, 210
Metal ions, desorption, 210
Metal-bound dimers, 212
Misalignments, 246
Mixture analysis, 99
Mode II, 166
Molecular beam, 173
Molecular ion, axial modulation,
    147
Molecular weight determination, 99
Momentum change, 79
MS/MS, see Tandem mass
    spectrometry,
    co-injection, 42
    efficiency, 111
    high mass resolution, 41–42, 132
MS/MS, of oligonucleotides, 134
MS/MS spectra, 100–101, 112, 115,
    117, 138–139, 231
MS/MS/MS, spectrum, 114, 133, 138
$(MS)^n$, 5, 100, 118, 139, 194, 197,
    approach, 101
    peptides, 111
    proteins, 111
    sequence, 101
    spectra, 100
Multipass optical arrangement,
    192–194, 199, 201
Multiphoton
    absorption, 164, 201
    collisional effects, 196
    dissociation, 165–168

infrared, 165–168, 192–199
    ion growth curve, 196
    ionization, 165, 173–174
    kinetics, 194
    mechanism, 194, 197
    processes, 165
protonated species, 166–167
wavelength dependence, 168
Multiple cleavages, 101
Multiplicity of peaks, 136
Multiply charged
    anions, 135
    biomolecules, 105, 108, 111, 124,
        127–128
    biopolymers, 92, 118–120
    even-electron ions, 131
    digonucleotide anions, 136–137, 139
    deoxyoligomers, 137
    peptides, 136
Multiply protonated
    molecules, 122
    peptides, 5, 129
    proteins, 129, 134
MS^3, 20–21, 100, 133, 137–138, 194
MS^4, 100
MS^5, 100

N

N-terminal amino group, 129
Negative ions, 82, 92, 134–135, 166,
        212
    anions, 135
multiply-charged, 135
    oligonucleotides, 137
    pulse detection circuit, 83
Neutrino, 257
Neutrino mass, 238
Nitrogen laser
Normalized emittance, 271
Numerical methods, 288
Numerical quadrupole simulation
        (NQS), 27

O

Oak Ridge ion trap system, 120
Oak Ridge National Laboratory, 93

Offset-DAC, 22, 26, 28
Oligonucleotides, 119, 134, 137
Oligonucleotides sequencing, 137,
        139
Online analysis, 111
Online separation, 116
Open cylinders, 248–249
Organic substrates, 210–211,
Orientational component, 153
Oscillation potential, 94

P

Particle collections' phase space, 265
Particle extraction, 291
    fast, 293
    slow, 292
Paul ion trap, 62, 251, 263–266,
        280–282, 284–285, 287, 290–292,
        294, 296, 298, 300
    collection device, 263
    ideal, 285, 287
Peak
    centroid, 146
    matching routine, 6, 39, 42
    shape, 107, 220
    spacings, 116
Penning ion trap, 62–63, 237–240,
        244, 247, 249, 251, 254–255,
        257, 260, 292
    general properties, 239
    ion trajectory, 241, 252
    mass spectrometer, 256, 259, 264
    practical realization, 243
    schematic diagram, 253
    tandem, 258
    tandem mass spectrometer, 258
Peptides, 116, 132
    doubly charged, 125
    high resolution MS/MS, 110
    hypothetical, 125
    linear, 124
    multiply protonated, 129
    sequence determination, 115
Personal computer, 220
Phase and phase angle, see RF,
Phase angle, 221, 223, 225
Phase-locked ion extraction, 298

Phase shift, 184–185
Phase space, 265–266, 279–280
  axial, 276
  condensation, 281
  distribution, 285, 287
dynamics, 275
ellipse, 71
  formalism, 265
transverse, 279
  volume, 276, 282–283
Phase synchronization, 183
Photo-absorption cross-section, 168,
  176, 201
Photodissociation, 163–166, 168, 170,
  173–174, 180, 189
  efficiency, 169–170, 196, 198–199
  products, 187
  rate, 168, 198
rate constant, 194
  reviews, 166
  spectroscopy, 174
  spectrum, 164
  threshold, 174, 176
UV, 171
UV-VIS, 169
Photoinduced dissociation (PID), 164
Photon absorption cross section, 166
Photon flux, 166–167
Photon-induced ion decay, 164
Photos, 165
Photoreaction channels, 165
PID, spatially resolved, 165
PID spectrum, 170
Polarizability, 145, 153–156, 158,
  196–197
Polarization
  atomic, 153
  electronic, 153–158
  frequency dependence, 154
  orientational, 153
Potassium ion affinity, 211, 121
Potential
  energy surface, 178,
  well, 278
  well depth, 180–181
Precise ion spacings, 44
Precision mass value, 257
Probe surface, alignment, 209

Product ion charge state, 132, 134
Programmable waveform generator,
  279
Proteins, 116, 132
  folding, 130
  high resolution MS/MS, 116
  multiply protonated, 129
Proton, 247
Proton-bound dimer ions, 121,
  123–124, 126–127, 129, 133,
  165–168
Proton transfer, 119, 121, 123–124,
  126–127, 130–134, 179
  intramolecular, 124
  reaction, 122, 130
Protonated
  molecular ion, 40, 167
  molecules, 40, 96, 101, 225, 229
Protonated species, 167–168
Pseudomolecular ion, 99
Pseudopotential well, 45, 291
  depth, 113, 128, 188, 284
Pulse peak power, 228
Pulse sequence, 82
Pulse shapes
  observed, 300
  simulated, 300
Pulsed
  ion excitation, 296,
  IR laser desorption, 207, 209
  operation , 55
  pressurization, 66
  trapping, 56
  trapping mode, 58
  valve, 67, 192–193, 197–198
Purdue University, 7, 27

Q

q_{eject}, 69–70, 73, 81, 83–84
q_z, 8, 10, 15, 35, 44, 58, 84, 112–113,
  147–149, 151, 185, 187, 225
q_z, definition, 10
Quadruply
  charged ion, 107
  protonated ion, 26, 33, 107, 117
  protonated Melittin, 131
  protonated Renin substrate, 117

Quadrupole field contribution, 188
Quadrupole ion trap, 4, 53, 57, 89,
        129, 139, 166, 180, 192, 205, 266
    cross-sectional view, 57
Quadrupole mass filter, 61, 66, 68,
        72, 106, 153, 166, 176
Quadrupole/quistor/quadrupole,
        69–70, 73, 81, 83–84
QUISTOR, 165–168

R

r_o, 8, 10
r_o definition, 10
Radiation damping, 251
Radionuclides, 264–265, 280
Raman spectrum, 176–177
Random noise, 103–104, 115
Rate constant, 84, 121
Reaction coordinate, 124
Reaction rate constants, 80, 84
Reduced scan rate, 108
Refractive index, 154
Relaxation, 165
    process, 76
    time, 168
REMPI, 175–177
Renin substrate, 33, 117
Resistive cooling, 250
Resolving power, 5
Resonance frequency, 16
Resonance ejection, 14–16, 94, 106
    frequency, 16
    scan, 108
Resonance-enhanced multiphoto
        ionization, see also REMPI,
        175
Resonance excitation, 5, 15, 94, 99,
        200, 246
Resonantly excited ions
    kinetic energy, 182
    position, 182
    velocity, 182
Retardation voltages, 280,
Reverse scanning technique, 11
Reverse-then-forward isolation, 20
RF
    amplitude, 94, 225

DAC steps, 22–23
    heating, 291
    ion trap, 64
    only mode, 15
    phase, 221, 225
        phase angle, 221–222
        phase dependency
        phase synchronization, 72, 223
    potential, 166, 174
    ramp speed, 12
    reduction, 12, 14
RF/DC isolation of precursor ions,
        5, 171
RF drive
    amplitude, 283
    frequency, 188
    frequency reduction, 7, 11, 12, 14
    frequency tunable, 18
RF drive voltage
    amplitude, 23, 113, 174, 209–210,
        294
    gradient, 113
    optimized, 7
Ring electrode, 11, 94, 166, 171, 245,
        278
    aperture, 187, 192
    axial slot, 180
    holes, 166, 207, 218–219
    hyperboloid profile, 11
    surface, 219
Runge-Kutta method, 284
Rydberg constant, 238, 256

S

Saturn II, 91
Scan function, 7, 94
    LDI, 219
Scan rate, 22, 107–108
    attenuated, 14
    reduced, 12, 22, 108
    slow, 5, 39, 43, 117
Scan speed, 107–109
Secondary ion mass spectrometry
        (SIMS), 5–7, 9, 11, 13
Sector instruments, 106, 217
Selected-ion/selected-molecule
        reactions, 69

Shimming coils, 246
Sideband cooling, 250
Signal enhancement, 95
Signal to noise ratio, 30, 108, 264
Silica optical fiber, 169
SIMION, 71, 73
Simple harmonic motion, 185,
    275–276, 287
Simple harmonic oscillator, 184,
Simulation studies, 26, 284
Single
    ion counting, 82
    ion detection, 255
    proton, 165
Singly charged positive ion, 95
Site of protonation, 124
Slow scan, 5, 39, 43, 117
Sodium
    adduct, 231
    exchange, 136
    ion incorporation, 135
Space charge, 44, 246
    effect, 150, 220, 287, 290
    forces, 291
    limit of saturation, 281–282
    tolerance, 30
Spatially-resolved PID, 165
Spectroscopy, 165, 174
Spherical coordinates, 243
SQUID, 254
Stability criterion, 242
Stability diagram, see also Stability
    region, 9, 16, 147
    boundaries, 9, 15
    for quadrupole ion trap, 16
Stability region, see also Stability
    diagram, 54, 75
Statistical mechanics, 286
Steel mesh, 173
Stochastic cooling, 251
Storage time, 57
Stretched geometry, 11, 91, 185, 188
Supercritical fluid chromatography,
    69
Surface-induced dissociation (SID),
    67
Synchronization , 81
Synchrotron ring, 281

T

Tandem mass spectrometry
    (MS/MS), 5–7, 20, 30, 41, 100,
    116, 119, 139, 217–218, 226, 228
    co-injection, 42
    efficiency, 111
    high resolution, 30, 32, 41–42, 116,
    118
    tandem in time, 139
Theory of quantum electrodynamics,
    259
Thermally labile sample, 217, 226
Thermospray, 5
Three-dimensional momentum
    space, 265
Three-dimensional space, 265
Tickle
    activation,152
    voltage of, 101, 151
Time
    domain data, 189, 290
    domain data, simulated, 191
    focus, 269
Time-of-flight (TOF), 269, 295,
    297
    detection, 256–257
    mass spectrometer, 4, 288
    method, 58, 218, 265
    profile, 254
    spectrum, 256
Timing diagram, 7, 18, 42
TIP mass spectrometry, 176–179
Trajectory
    amplitude, 244
    probe, 180
Transformation of axial energy, 282
Transition metal ions, 212–213
Transition metal ions, desorption,
    213
Transverse action space, 280
Transverse phase space, 279
Trapped ion photodissociation, TIP,
    175
Trapped ion signal intensity, 74–76
Trapping
    efficiencies, 64–65, 79, 85, 111,
    220–221, 224

field imperfections, 244
  process, 63
Trigger circuit, 222
Triple stage quadrupole instrument, 228
Triply
  charged ion, 117
  protonated Melittin, 131
  protonated molecule, 31, 107
Two-dimensional energy diagram, 125, 127
Two-photon ionization, see also TIP, 175

U

Uniform force, 270
Unimolecular reaction, 128
University of Mainz, 243
University of Washington, 243
Unstable isotopes, 259

UV-absorbing matrix, 227
UV photodissociation, 171
UV-VIS, 165

V

Varian 3400 gas chromatograph, 171
Varian Saturn II, 91

W

Wire mesh, 57, 173
Wronskian determinant, 36

Z

z-oscillation amplitude, 246
z_o, 10
z_o, definition, 10
Zoom, 107, 117, 135
Zoom scan, 117